Chasing Science at Sea

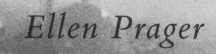

Chasing Science at Sea

RACING HURRICANES,

STALKING SHARKS,

AND LIVING UNDERSEA

WITH OCEAN EXPERTS

THE UNIVERSITY OF CHICAGO PRESS

CHICAGO AND LONDON

Ellen J. Prager is currently the chief scientist at the world's only undersea research station, *Aquarius Reef Base,* in the Florida Keys and a freelance writer. Among her publications are *The Oceans and Furious Earth: The Science and Nature of Earthquakes, Volcanoes, and Tsunamis*; a series of children's books including *Sand, Volcano,* and *Earthquakes* with the National Geographic Society; and a children's novel, *Adventure on Dolphin Island.*

The University of Chicago Press, Chicago 60637
The University of Chicago Press, Ltd., London
© 2008 by Ellen Prager
All rights reserved. Published 2008
Printed in the United States of America
17 16 15 14 13 12 11 10 09 08   1 2 3 4 5

ISBN-13: 978-0-226-67870-2 (cloth)
ISBN-10: 0-226-67870-9 (cloth)

Library of Congress Cataloging-in-Publication Data
Prager, Ellen J.
  Chasing science at sea: racing hurricanes, stalking sharks, and living undersea with ocean experts / Ellen Prager.
        p.   cm.
  Includes bibliographical references.
  ISBN-13: 978-0-226-67870-2 (cloth: alk. paper)
  ISBN-10: 0-226-67870-9 (cloth: alk. paper) 1. Oceanography. 2. Ocean. I. Title.
  GC11.2P738   2008
  551.46—dc22
                                    2007049486

To my parents,

for encouraging my explorations

of nature as a child and for fostering

my never-ending curiosity

CONTENTS

ACKNOWLEDGMENTS

My sincerest thanks go to all of my colleagues, too many to name here, who were willing to share their field experiences for this book and endured my numerous phone calls and e-mails. Your insightful contributions transform the book from a personal account into a wide-reaching, entertaining, and powerful look at the benefits of doing fieldwork that spans the disciplines of ocean science. Special thanks to Linda Glover, Bob Halley, Gene Shinn, and Robin and Jan Hawk for your encouragement and for always providing thought-provoking conversation or at least a good laugh over cocktails. Thanks to my sister, Kathy, and her family for their support and for providing wonderful meals as well as good company for a wayward aunt on her travels. My appreciation also to the publisher, the University of Chicago Press, and especially executive editor Christie Henry for her enthusiasm for the project and her excellent editorial skills; when the sentences or words just didn't look right, she constantly brought fresh eyes and sage advice to the page. Thanks also to Joann Hoy, whose editing greatly improved the flow of the text.

My deepest gratitude goes to my sponsors, the University of Miami's Rosenstiel School of Marine and Atmospheric Science, the National Marine Sanctuary Foundation, and the Wildlife Foundation of Florida. Their financial support was invaluable and allowed me to dedicate my time and attention to the book. Special thanks to Dean Otis Brown for his continued support of my nontraditional efforts to bring ocean science to the public. Appreciation also to Dan Basta and Matt Stout at the National Marine Sanctuary Program; Lori Arguelles at the National Marine Sanctuary Foundation; Wendy Quigley, Gil McCrae, and Ken Haddad at the Florida Fish and Wildlife Research Institute and Conservation Commission; and all of my other friends and colleagues who helped make this book a reality.

AUTHOR'S NOTE

~~~~~~~~~~~~~~~~ My colleagues and I did our best to portray the events recounted in this book as accurately as possible; however, there may still be slight biases created by the working of time and age. I personally apologize to anyone who remembers such incidents differently or is, in any way, offended by our recollections. The stories in the book are a mere fraction of the wonderful adventures and experiences of people who have dedicated their lives to the study of the ocean. The book is not meant to be a comprehensive overview of doing science at sea, but rather a way to begin a dialogue that uses storytelling to share with others our passion for and understanding of the ocean and science.

PROLOGUE

～～～～～～～～～～ An infamous shipwreck finally found, an elusive, strange ocean creature seen for the first time, or a fearsome attack by a powerful seagoing predator. These are the ocean stories we see or hear about, the ones that make it on to the television, Internet, or pages of a magazine. Yet they are merely specks within the tapestry of our knowledge about the sea gained over centuries of exploration and study. The true fabric of our understanding comes from the cumulative efforts of scientists, mariners, explorers, and others who spend their time, often their lives, going to sea and trying to unravel its mysteries. Their efforts, day after day, week after week, reveal the true nature and wonder of the ocean along with its critical role on Earth. They also illustrate the realities of how science is done. For many ocean scientists, myself included, the sea is a passion; some might say an obsession.

Only a certain breed chooses to go to sea for weeks at a time without the comforts of home, family, or even friends. To go into the field, one must be willing to face the risks of severe weather or unpredictable marine life, to endure physical and mental discomfort, and to suffer the frustrations of mechanical failures or simply bad luck. When one is spending extensive time on or under the ocean, a good sense of humor comes in handy, or maybe just an antisocial personality. The chance to observe and learn about the ocean firsthand can, however, be an irresistible draw. For me, and many of my colleagues, it is why we became ocean scientists. And while going into the field is a critical part of our research in the sea, it also provides us with some incredible tales to tell, stories of adventure, wonder, humor, inspiration, and near-disaster. From each of our experiences, we have learned about the ocean, science, and ourselves.

Stories from the sea also make for entertaining conversation, whether over a steaming cup of coffee or a glass of fine champagne. My own sea stories have amused audiences for years, especially at cocktail parties (possibly assisted by the alcohol consumed by the audience as well as the storyteller), from working as a support diver—a euphemism for underwater slave—to living undersea for weeks at a time, taking students to sea aboard a tall sailing ship, or going on research expeditions to exotic locales. I've encountered equipment-stealing sea lions in the Galápagos, worked with ex-NFL football players turned underwater shark-wrestling stuntmen, nearly capsized on a trawler while entering a dangerous inlet, faced a hurricane at sea with a boat full of undergraduates, and stood waist-deep in steaming mud as turkey vultures circled overhead. My many field-going colleagues have stories equally entertaining and just as enlightening. They are an eclectic group that includes marine geologists, biologists, physical oceanographers, marine engineers, expedition doctors, ship captains, and deep-sea divers. They are often an irascible and cynical bunch, yet also dependable, funny, surprising, and, at times, brilliant. And while we laugh at our mishaps or ponder some of the strange things we've seen in the sea, we are continually inspired to learn, driven by an insatiable curiosity.

In ocean science today, research is increasingly reliant on remote technologies, such as satellite imaging, robotics, and computer modeling. Funding to support fieldwork is decreasing, and society is increasingly adverse to the risks involved. Field-based science is

becoming less prevalent and harder to undertake. And for those who
were pioneers in the field, investigating topics such as coral reefs or
the deep sea, time is wearing on. The knowledge they gained from
days spent at sea and their inspirational tales of adventure may soon
be lost forever. I have therefore endeavored to write this book as a
means to share our stories, illustrate the importance of fieldwork,
and provide a behind-the-scenes look at how science really works.
The following pages are not meant to be about my colleagues or
me, but rather about what we have experienced and learned from
doing fieldwork in the ocean.

1 *An Introduction*

FOR CENTURIES, EXPLORERS AND SCIEN-
tists have taken to the high seas or journeyed into
the ocean's depths to view, measure, and study its
wonders. The methods they use range from a simple bucket
swung overboard to a highly sophisticated, remotely operated
vehicle towed miles beneath the surface. As modern ocean scientists,
my colleagues and I now have an amazing array of technology at
our fingertips. From space we can get a snapshot of the entire sea
surface at one time, seeing the ocean from a larger perspective than
ever before. Meanwhile, molecular technology is enabling us, for
the first time, to study in detail the smallest creatures of the sea, the
microbes, and examine their role in the overall ocean ecosystem.
Genetics are revealing connections between marine populations
once thought separate and showing that other organisms have
distinct hereditary histories. Remotely operated undersea vehicles
(ROVs) are providing unprecedented access to the deep sea, while
new autonomous vehicles can be programmed to explore under the
ice and in other inaccessible or dangerous environments. With tags
tracked via Earth-orbiting satellites or through undersea acoustics,
we can now follow the sea's creatures beneath the waves and around
the world. But even with such great advances in technology, we

must still go into the field to effectively study and understand the ocean.

Technology alone does not avail all we need to know about the sea. The information provided by satellite imagery or computer simulations is literally worthless without real-world observations for input and verification. In the laboratory, we can create and run complex experiments, but we still need to go into the field to determine if the results are realistic and applicable in nature. To tag marine organisms for tracking or to observe their natural behavior, we must go to sea. We need to go into the field to sample sediments, to extract cores for geological study, and to collect water for chemical or biological analyses. In the field, we deploy drifting buoys to track ocean currents, maintain observation platforms and sensors, and monitor long-term change. But going into the field is not just about the planned collection of data or the deployment of equipment; it is also about the unplanned and serendipity. With time spent in, on, or under the ocean come revelations that can prove our previous assumptions wrong, stimulate new lines of investigation, and provide great moments of wonder that inspire a lasting sense of discovery. And as much as going into the field can teach us about the ocean, it also unveils how we learn and the true workings of science.

DOING SCIENCE

In school as well as in many textbooks, the scientific method is routinely described as a rather boring, orderly process by which hypotheses are created and then tested. Not a very inspiring notion or one that readily draws students into a related career. At its core, science is simply about observing and then making sense of those observations to better understand the world around us. In contrast to the stereotype, this can be an exciting, challenging process that involves humor and creativity and that evolves with time, taking as many twists and turns as a compelling drama or good suspense novel.

Doing science is not about following a cookbook-type procedure; it is about curiosity, questioning, and encountering the unforeseen and unplanned. For many scientists, myself included, what we enjoy most about science is the constant learning involved and the frequent surprises that come with the pursuit of knowledge, especially when working in the field. I can think of no better way to illustrate the joy, frustration, and complexities involved in science than to take a behind-the-scenes look at ocean science, starting with a brief introduction and a few stories from the field.

ASKING A GOOD QUESTION

Most science begins with a question. Bob Ginsburg of the University of Miami's Rosenstiel School of Marine and Atmospheric Science is a world-renowned coral reef geologist and famous for his uncanny ability to ask really great science questions. At a presentation's end, even when I swear he was sleeping throughout, he consistently asks the one question we are all thinking of or wish we were thinking of. He has been known to stand up and in a deep commanding voice simply ask, "So what?" Graduate students fear his insightful questions, and scientists, if not prepared, can be caught off guard. He teaches us all one of the most critical parts of doing science, being able to ask a good, provocative, and relevant question.

In ocean science, questions can be as simple as what geological features or organisms are found at the seafloor in a certain location, or they can be more complex, such as what is the role of the ocean in climate change. Where do research questions come from? In applied science, research is undertaken to address a problem with direct societal impact. In the ocean, examples of applied science include research on harmful algal blooms, invasive species, tsunamis, the ocean's role in global warming and hurricanes, coastal pollution or beach erosion, and fisheries. When research is done to answer a question or gain understanding that is not directly identified with a societal problem, it is known as basic science. Examples of basic science in the ocean include the exploration of deep-sea environments or research on marine mammals, sharks, the biology of corals, and the overall productivity of the ocean. Basic research, however, frequently leads to understanding that can be applied to societal issues. For instance, some bacteria discovered in the deep sea can grow using substances that we consider toxic, while others are able to produce electricity from the surrounding seawater. Scientists are exploring how these microbes might be used for toxic-waste cleanup or to generate power for undersea instrumentation. Development of technology for basic research can also prove beneficial to society.

Fieldwork itself, more often than not, leads to new research questions or areas of investigation. Coming up with more questions than answers can be frustrating, but this is an important part of the process of doing science. For me, this is one of the best parts of doing fieldwork and one of the reasons I have particularly enjoyed working in Florida Bay. It is not the most scenic place I've studied in nor does it host many of the ocean's more charismatic creatures, but nearly every excursion into Florida Bay reveals something new or unexpected. As a researcher with the U.S. Geological Survey in the 1990s, I spent months in Florida Bay studying sediment and wave dynamics with my colleague and good friend, marine geologist Bob Halley.

Florida Bay is the shallow, wedge-shaped region that sits between the southern extent of the Everglades and the western edge of the Florida Keys. Along with its importance environmentally, Florida Bay provides a significant source of economic revenue for the region through recreational and fishing activities. Since the late 1980s, Florida Bay has been under intense scientific scrutiny due to a massive seagrass die-off and subsequent algal blooms, episodic sponge mortalities, and decreases in water clarity. With restoration in mind, over the past several decades, researchers from a myriad of disciplines, agencies, and institutions have focused their efforts on trying to better understand the bay as well as the events that have occurred there.

The first lesson from fieldwork in Florida Bay—it probably shouldn't be called a bay. The area is actually a series of small, shallow embayments separated by an interconnecting network of mudbanks and mangrove islands. Many a boater has become personally and rudely acquainted with these mudbanks, as they can be difficult to see and have the tendency to reach out and grab small boats—or so it would seem. Park rangers patrol Florida Bay, as it is part of Everglades National Park. One ranger tells of a weekend boater who, after running hard aground on a mudbank, proclaimed his innocence because the chart he was using showed that he had been in sufficiently deep water—the problem was probably that his chart was actually a "maplike" place mat he had picked up at a local restaurant. Lesson number two when doing fieldwork in the ocean—Florida Bay or elsewhere—bring along an accurate chart. But truthfully, even with the best of charts, Florida Bay can be difficult to navigate, especially when strong winds stir up the bay's fine carbonate sediments, turning the water the color of a milk shake.

Bob and I spent weeks in Florida Bay mapping seafloor habitats, sampling sediments, and conducting experiments. Much of the data we collected was as expected, but almost every day we would also observe some new phenomenon or see something intriguing. We saw a surprisingly aggressive ray chasing a shark, a rare baby crocodile in the mangroves, and powerful squalls that seemed to come out of nowhere. And just when we thought we had a good handle on the basic sediment, seagrass, and water patterns in the bay, we were once again surprised.

One day while snorkeling in one of the region's small bays, we discovered a surprisingly warm layer of water at the bottom. Based on our understanding of Florida Bay and simple physics, this should not have been the case—or so we assumed. In Florida Bay and elsewhere, warm, therefore less dense, seawater typically resides at the surface above cooler, denser seawater. By afternoon in the tropics, due to the sun's heating, anyone swimming, snorkeling, or diving in shallow water can observe firsthand the development of a warm layer of water at the surface. But rarely do you find warm water at the bottom—the density thing—and this certainly did not jibe with our understanding of water flow in the bay. We had no way to measure it at the time, but the only explanation we could come up with was that the warm water at the bottom was higher in salinity than the overlying water, thus making it denser. How would the salinity of the bottom water increase? Our first thought was that it could be groundwater flowing up from an underground spring—common in the limestone underpinnings of southern Florida—but then it would have been cool and less salty. We continued to ponder the question until later that day when we nonchalantly jumped out of our small boat onto a mudbank.

The water on the crest of the mudbank, only a few inches deep, was scalding hot! Bob and I made a hasty retreat back into the boat and while nursing our scorched toes came up with a theory. During hot, sunny summer days in Florida Bay, the shallow water on top of the mudbanks heats up and is subject to high evaporation, thereby increasing its salinity enough so that the water becomes denser and flows off the mudbank into an adjacent bay, settling at the bottom. We had heard of this density cascading effect in other ocean environments.

While the incongruously warm water at the bottom in Florida Bay was not the main thrust or even part of our research, it was too intriguing for us to ignore. One day, in between our official studies, we used a small temperature and salinity probe to test the warm water at the bottom of a small embayment. Sure enough, it was higher in salinity than the overlying seawater, thus supporting our density cascading theory for its origin. We never had a chance to do a thorough investigation of the phenomenon, but since then other researchers in Florida Bay have mentioned finding surprisingly warm water at the bottom. We also wondered what the impact of this hot, salty water was on the marine organisms living in the bay's seagrass and sediment.

Florida Bay, like many marine environments, is highly complex, and even with years of study we are just beginning to understand it. For the curious, this is part of the fun of science and especially of working in the field.

I cannot resist the opportunity here to recount one of the most famous stories about how unexpected observations in the field led to new science questions and, in this case, to an entire revolution in our way of thinking about the deep sea. In February 1977 a team of scientists embarked on a cruise to study the Galápagos rift, an area of seafloor spreading some 400 miles (644 km) west of Ecuador and 250 miles (402 km) northeast of the Galápagos Islands. They were investigating the possible presence of hydrothermal vents, at the time a relatively new concept describing fractures in the seafloor, thousands of feet below the ocean surface, that emit warm chemical- and mineral-rich water.

Once on site, the team of geologists, geochemists, and geophysicists deployed *Angus*, an ROV, to investigate the rift. With its specialized deep-sea cameras and temperature sensors, *Angus* was towed just above the seafloor in depths of over 8,000 feet (2,438 m). A titillating spike in water temperature was detected, possibly indicative of a hydrothermal vent, and when photographs taken of the seafloor were developed, an even greater surprise was revealed—clusters of giant clams and hundreds of mussels. Up until this time, the deep sea, far removed from the life-giving powers of the sun, was thought to be much like a desert, devoid of plentiful marine life.

The deep-diving submersible *Alvin* then arrived on the scene aboard its support ship, the R/V *Atlantis* out of the Woods Hole Oceanographic Institution. Two scientists and a pilot squeezed aboard *Alvin* to investigate the possible vent and the strange creatures at the seafloor: little did they know their journey would be one for the history books. Guided by the data provided by *Angus*, the sub driver steered the undersea vehicle down toward the area of warm water. As they began to get a firsthand look at the seafloor, one scientist reportedly questioned the notion of the deep sea as a desert—because outside the viewport there was an oasis of life. It was an abundance of organisms never before seen at such great depths, and many of the creatures were completely new to human eyes. Along with large clams, the researchers saw white crabs, a purple octopus, and lush gardens of strange foot-long tube worms with white stalks that were topped by bright red, plumelike gills. There wasn't even a biologist on site because the organizers of the expedition didn't expect to find marine life in the rift zone. Rumor has it that when the geologists aboard reported their findings to biologists back on shore, they were highly skeptical, to put it mildly. The team had discovered a previously unknown, even unimagined, ecosystem in the deep sea where organisms thrive on the chemicals and bacteria associated with hydrothermal vents. From this one discovery, a myriad of new scientific questions arose, including whether the deep sea is where life on Earth began, and it created several entirely new fields of research in ocean science. Scientists now estimate that there are some 280 active vent sites within the world's oceans—most of which remain unexplored!

THE MONEY

Along with even the best of questions, ocean science, unfortunately, takes money and usually lots of it—ocean research isn't cheap. Time at sea aboard a large research vessel can cost tens of thousands of dollars per day, and even small-boat work takes a fair bit of cash—especially with today's fuel prices. Even the earliest explorers of the sea needed investors and private sponsors to support their voyages of discovery. Today most scientists rely on grants from the government, foundations, or private institutions. Sadly, however, in recent years ocean research and exploration have not been a national priority. This is especially frustrating given our connections to and reliance on the ocean as well as the changes we're causing in the sea. Whereas we spend approximately $17 billion on space exploration and research each year, the nation annually invests only about $700 million on the science of the oceans, which, by the way, cover nearly three-quarters of the earth's surface. Experts estimate that we have explored only about 5 percent of the sea.

With funding limited and the needs great, the competition for grants in ocean science is fierce. Researchers must constantly write proposals that enumerate their expertise and knowledge, that prove they have access to the necessary equipment, adequate personnel, and enough preliminary data to justify the research for which a grant is being sought—the latter presenting a conundrum of sorts. A researcher needs sufficient data to get funding to collect data.

Finding funding sometimes seems like a lifelong pursuit in science—beginning in college and progressing through and after graduate school. When considering science as a career, students often ask and worry about the cost of graduate school. Luckily, there are many options for financial assistance for graduate training. My experience demonstrates both traditional and less conventional means of support. For my doctorate degree, I was fortunate to obtain a wonderful fellowship at the Coastal Studies Institute at Louisiana State University in Baton Rouge. However, to pay the bills while studying for my master's degree at the University of Miami's Rosenstiel School, I worked as a graduate research assistant, took out a loan, and supplemented my meager income through a variety of interesting field-based jobs.

Prior to entering graduate school, I acquired a strong background in fieldwork and science diving. For example, during the summer between my junior and senior year as an undergraduate at Wesleyan University, I worked as a support diver for *Hydrolab*, a small undersea research laboratory operating on the northwest coast of St. Croix in the U.S. Virgin Islands. The other support divers and I were the undersea gofers. We were responsible for filling and transporting scuba tanks, tracking scientists when they were diving outside of the undersea station, shuttling food and equipment daily to and from the laboratory, cleaning the habitat before and after each mission, and assisting elsewhere as needed (more in chapter 6). It was the best summer job ever! With this and other field experience under my belt, I was able to pick up several short-term jobs in Miami to help pay the bills. I helped supervise diving on a project off the coast of New Jersey to test new seismic equipment. It was cold and dark, there were strong currents, and I think the scientist in charge of the project was crazy—literally, let alone completely unaware of the issues of working at sea. For instance, when an instrument is attached to a ship by a cable and the ship swings, without slack in the cable, the equipment and any divers working on it will swing as well and be dragged over and through the sand and mud. A high concentration of sediment in the water often led to diving by feel—not a practice I highly recommend. Another of my jobs was planting seagrass for a female scientist who clearly had a preference for strapping young male divers—I didn't last long on that one. But my favorite job in graduate school was when I went to the Bahamas with two ex-NFL football players turned stuntmen turned fish collectors for a public aquarium. Our voyage was one adventure after another.

Early in the trip, while caught in a bad storm, we spent the night looking for a safe anchorage, all the while trying to keep the small boat we were towing afloat. The skiff eventually flipped and was lost, though only temporarily, as we got it back the next day after bartering for its return with some locals who found it washed ashore. We did, however, have one moment of late-night humor—at least I did. As we were tending the small boat off the stern, a flying fish leapt out of a large wave and, much like a pie-in-the-face gag, flapped directly into the face of one of the other divers. Flying fish are amazing creatures. They jump out of the sea, extend their fins like wings, and use their tails as a rudder to skim gracefully over the surface. This behavior is believed to be an escape response to avoid potential predators. The waves that night were so large that when the flying fish leapt out of the sea, it was already at eye level and perfectly aligned for an in-the-face landing, providing at least one of us onboard with some comic relief. A colleague of mine was once asleep aboard his sailboat when a flying fish leapt in through an open porthole and landed, still flapping, on his chest—talk about a rude awakening.

Back to the Bahamas in the storm. Very early the next morning we finally anchored and hit our bunks for some much-needed sleep. A few hours later, however, we all awoke to a rather unsettling, unnatural roll of the ship—no roll at all. The anchor had dragged, and the boat was now sitting high and dry perched atop a shallow bank. All we could do was wait for the tide to come up and float us off. Another lesson of fieldwork: make sure your anchor is firmly and securely set. With shifting winds, waves, and currents, let alone variable bottom types, the ocean can be a tricky place to anchor. I once watched a very experienced dive team surface only to find that their boat had disappeared. Luckily, while diving nearby, my buddy and I had seen their anchor and its chain float by—literally. We grabbed the anchor and secured the boat next to ours. Of course we couldn't help but wait with great anticipation for the surprised look on the other divers' faces when they surfaced without a boat. It was all in good fun, since we had their boat and they were safe and unhurt, though a bit embarrassed.

Once again, back to my Bahamian fish-collecting adventure—I mean income-generating job for graduate school. On another morning we awoke to find a surprise in one of the wells where we stored our live fish. There was a large barracuda in the tank, now fat and happy, having eaten much of our previous days' work. The captain demanded to know which of us had been stupid enough to put the barracuda in with the other fish. All of us vehemently denied putting the fish in the tank as well as being that stupid. We then discovered fish scales on the rail surrounding the live well, leading us to surmise that the barracuda had actually jumped aboard sometime during the night—either a very smart or very lucky fish. My companions on the cruise later offered to show me how to wrestle sharks as they had done as stuntmen for a previous Bond movie—they also informed me that after being drugged and put in the hold of a fake ship, sharks, once they come to, are none too happy! I decided that shark wrestling might not be conducive to finishing my degree or a life skill that I really needed.

My point here is that finding funding for research or graduate study is part of doing science and there are options out there, albeit some more conventional than others. As for research grants, I won't go into the agonizing process of writing and submitting a proposal or the exorbitant amount of time a scientist must now spend doing this. Nor will I discuss the bureaucracy involved and how red tape and administrative duties can suck all the joy out of science and even fieldwork. But once funding has "somehow" been obtained from "somewhere," scientists then plan and execute their study—an exciting, challenging task when working at, on, or under the sea.

THE LOGISTICS OF FIELDWORK

For the uninitiated, spending days doing research while cruising aboard a ship or living on a remote tropical island sounds glamorous, a vacation of sorts. Glamour rarely comes into it. Working in the field typically entails few, if any, of the creature comforts we associate with cruise ships or tropical island getaways. And the planning for fieldwork is definitely more than simply packing a suitcase. Success in the field takes organization, patience, perseverance, and usually hefty doses of creativity and good luck. The logistical difficulties of working in the field can in fact winnow out the serious oceangoing researchers from would-be adventurers.

Organizing for fieldwork at sea is usually about two things, taking as much as possible to make your research successful in terms of equipment and supplies and bringing along as little as possible for yourself, personally. Scientists need to be prepared for what could, and most likely will, go wrong with their equipment or their research plan, so they need to bring along plenty of spares and adequate repair supplies. On the other hand, since traveling can be difficult and the living quarters aboard ships and at field stations are minimal, clothes and personal items must be kept to the bare essentials.

A narrow shelf next to your bunk is all the storage space for clothes and personal effects one may have while in the field. Such minimalist living may seem untenable to most people, but it's quite doable, especially for relatively short research ventures. For longer expeditions or voyages, doing laundry is a good thing. What passes for laundry, however, can be questionable. While teaching oceanography aboard Sea Education Association's 130-foot-tall sailing ships, I went to sea for six weeks at a time with some twenty-five undergraduates and ten crew. As chief scientist my quarters were in the aft cabin, shared with the captain. We had relatively spacious accommodations, each having a large bunk, wide shelf, and a few drawers for clothing. Others onboard, especially the students, had little more than a bunk the likes of which remind one of an oversize coffin. My bunk was large enough so that clothing stuffed along the sides was essential in rough weather to prevent rolling and some serious head banging. Even with the luxury of drawer space, doing laundry was desirable after several weeks at sea. Since freshwater was in limited supply, seawater was used for washing—a process that leaves clothes oddly stiff, yet strangely soggy. After weeks at sea, that first truly laundered shirt just smells and feels oh-so-good—to say nothing of the freshly showered people. The lack of creature comforts in the field makes one more appreciative of the simple, everyday things we so often take for granted. After two months of research in the Galápagos Islands in the late 1980s, a cheap hotel room in Ecuador with a hot, freshwater shower and a soft bed without spiders, anchors banging overhead, or thieves (more in chapter 5) was the equivalent of a luxury five-star hotel.

Time spent in the field can also make a person truly appreciative of a home-cooked meal, a stationary dining table, or simply access to a grocery store with fully stocked shelves. After weeks of eating food of questionable origin in the Galápagos, a jar of peanut butter and a container of oatmeal kindly left behind by a group of students were like manna from heaven. From the art of dining at a tilting gimballed table in rolling seas to the oddities of meals in a high-pressure undersea habitat or while in an exotic port of call, food may go beyond being a necessity and can become a source of amusement, comfort, or, sadly, misery. Knees and elbows can be a real problem when eating at a gimballed table. If a person mistakenly stops the table from tilting in rolling seas, meals slide oddly up the slant of the table or can be quite efficiently catapulted. At one Thanksgiving feast aboard a Sea Education Association (SEA) ship, a large turkey made a surprise postmortem flight into a student's bunk—a gimballed-table mishap. Mealtime in the field often provides humor as well as a welcome relief from the stress of the work involved, from being away from friends and family, and is an opportunity to bond with colleagues.

Logistical planning for fieldwork in the ocean must also take into account the operational setting. On a ship your entire work area may be pitching, swaying, vibrating, or slamming up and down. A rolling sea necessitates a means to secure equipment—or even people—in one place. In high seas on SEA ships, we sometimes rigged bungee-cord seat belts while working in the small laboratory. Some pieces of equipment, to say nothing of some people, just don't do well when vertical and horizontal stability are an issue. I'm always amazed by my colleagues who go to sea for a living, yet are prone to seasickness. And of course if you are actually working in the ocean, there are a whole host of other things to consider—especially the seawater itself. Besides being wet, seawater is extremely corrosive and can instantaneously suck the life out of electronics. The ocean also has a way of swallowing up expensive pieces of equipment, snapping or ensnaring cables, and causing guidance systems to fail. Boats and their engines require constant care and maintenance. Even the simplest of low-tech equipment when lost or broken can set a project back. When surveying a coral reef on scuba, we often use small pencils to write on underwater slates, yet they have a penchant for floating away. Researchers now have a wide array of creative ways to attach pencils to their slates, but somehow, someway, those pencils still find a way to escape or break. An early lesson in underwater survey work—always bring spare pencils.

The list of gear that is needed for research at sea can be extensive. How to store and secure all of the needed equipment, particularly on small boats, is truly an art form. More than one boat has left the dock precariously loaded. To make matters worse, some scientists have little experience operating and maintaining small boats or even working on large ships, yet their research requires these skills. It can be an interesting learning process. In the Bahamas I once came upon a graduate student enthusiastically loading a small boat with gear for his project. As he piled in equipment, the boat was quickly filling with water—it was already at his shins. I asked if maybe he had forgotten something important—like putting in the boat's plug (a small rubber stopper many small boats have at the base of the stern to drain water while at speed or when raised out of the water). Making sure the plug is in before loading a small boat can prevent an embarrassing incident at the dock—inadvertent sinking.

When working in the field and especially when traveling to

remote regions, one must also consider potential medical and safety needs, fuel, and communications. Safety has to be the top priority in any field outing, and precautions to prevent injury and sickness are critical. It is important to plan for possible emergencies and determine where and how to get medical attention if needed. Marine scientist Duane De Freese of Hubbs-SeaWorld Research Institute points out that, along with assuring the safety of the people involved in fieldwork, scientists must also consider the safety of any of the animals they are studying.

Just getting things to the right place at the right time can be easier said than done. I had the good fortune to accompany one of my professors in graduate school, Steve Murray, to Papua New Guinea for an oceanographic cruise. We were to meet the University of Hawaii's research vessel, the *Moana Wave*, in Papua New Guinea, load our equipment, and then spend a week or so deploying monitoring buoys and measuring ocean currents in an area where flow was poorly understood. From the beginning our plans seemed to go amiss. Travel delays resulted in a young graduate student—me—traveling by myself through Australia to Port Moresby (which at the time was reportedly quite a dangerous place, especially for a young, blond female) and on to Madang, on the northern coast of Papua New Guinea. The turbulent flight aboard a small plane from Port Moresby to Madang was made more memorable by the completely unintelligible announcements (I'm pretty sure they were in English) and the in-flight snack, which was served with great pride—a small paper cup of a mildly flavored orange drink accompanied by a miniature Spam sandwich. Spam is surprisingly popular in some areas of the world. Landing in Madang, I had no idea where to go or how to get there (another lesson learned). Luckily, a lovely local family offered me, and my gear, a ride to the marine laboratory in the back of their pickup—of course there was only one marine laboratory in Madang. Steve joined me within a day or two. Unfortunately, the same could not be said for his equipment.

The *Moana Wave*, a 200-foot research ship, arrived in Papua New Guinea right on schedule, but our equipment somehow ended up in Hong Kong. Large research vessels, such as the *Moana Wave*, are tightly scheduled for scientific cruises throughout the year and can't accommodate extended delays. It was either drastically alter our research plan or return home empty-handed, not knowing when or if the cruise could be rescheduled. Given the amount of time it takes to write a proposal, get funding, and then garner and schedule ship time, it had probably taken Steve two years or more to plan the cruise. Yet, while greatly disappointed, he took the situation in stride, grateful that although it was a serious setback to his research, no one had been hurt. We couldn't deploy moorings for long-term oceanographic observations, but all was not lost as we could still use the ship's acoustic Doppler system to measure ocean currents while transiting. Makes losing luggage on an airplane flight seem rather inconsequential. In planning for fieldwork, flexibility is an advantage, but once in the field a can-do attitude is a true necessity because, even with the best of plans, things can and do go wrong. It is all part of the challenge and the reward of doing science at sea.

INTO THE FIELD

The real adventure, learning, and fun begin after the background research is done, proposals have been written and funded, and the requisite planning is complete, because then it is time to go into the field. Doing science at sea—whether on a large ship, in a small boat, or aboard a deep-sea submersible—is always an adventure, and it often entails extended periods of boredom interspersed with intervals of extreme physical exertion, incredible excitement, and moments of inspiration. In the field, frustrations are forgotten as some of the ocean's most wondrous or mysterious sights are encountered (chapter 2). Serendipity and unexpected results often lead to profound insights (chapter 3). And when we're caught in a storm, being in the field makes us all forever more aware and respectful of the true power of the ocean and atmosphere (chapter 4).

Success at sea is amazingly rewarding, given the challenges faced. In the process of overcoming obstacles, we grow stronger and more skilled, and sometimes make technological advances or gain new understanding (chapter 5). Within the ocean, depth becomes our enemy as we seek time to observe and study the sea (chapters 6 and 7). While in the field, a scientist's flexibility and creativity are tested, and if nothing else, each of us learns to laugh at ourselves and the predicaments we get into. To illustrate this last point, I can't help but turn back to Florida Bay.

Bob Halley and I continually sought new and more efficient ways to walk and work on top of Florida Bay's mudbanks, without the usual sinking knee- to thigh-deep. The mud is especially soft and squishy due to its composition, made principally of fine grains of calcium carbonate. Trudging through the mud wasn't so bad where it was smooth and almost spa-like—except maybe for the thick, foot-long slithering worms found living there. But where the mud also contained an abundance of sharp, angular shell fragments, slogging through it was not only tiring, especially when carrying equipment; it was also quite painful.

We discovered that the national park rangers working in Florida Bay had found an ingenious way to walk on top of the mudbanks and avoid damaging seagrass—plastic snowshoes. While organizing our gear for work in the bay, I eagerly ordered a pair. Once in the field, Bob graciously allowed me to be the first to try the snowshoes out—or maybe he just knew better. I strapped those babies on and strode confidently across the seagrass. They worked great—until I hit a mud hole.

On Florida Bay mudbanks, there is extensive seagrass growth, yet there are also mysteriously open, round holes filled with extremely soft mud. I don't know exactly how they form (some people believe they are started by stingrays digging for food), but I do know, now, what happens when you step into a mud hole while wearing plastic snowshoes. Mud instantly gushes up through the gaps in the webbing of the snowshoes, and the person wearing them is instantly stuck in the mud in the hole. If that person happens to have been striding overconfidently forward, the result is an award-winning face-plant into the mud. Not my most graceful move, but one that afforded both Bob and me a good laugh. We reverted to suffering the heat in our wet suits to protect our legs and to some rather undignified crawling. After I left the U.S. Geological Survey, Bob continued his research in Florida Bay. I hear he discovered that when there are a few inches of water over a mudbank, a buoyant kickboard strapped to one's stomach with duct tape works well to skim across the top—I've seen photos. Ingenuity and a healthy sense of humor definitely contribute to the success and sanity of field-going scientists.

Being in the field also avails opportunity for invaluable experiential learning. For all the time we spend studying in a classroom or reading in our offices and the library, nothing can match the understanding we gain by simply being in the field and observing nature firsthand. In any basic oceanography course, students learn about how temperature changes with depth in the sea. Data from oceanographic cruises are used to illustrate the concept of a thermocline—where relatively large temperature changes occur over a short range in depth. We can read or teach about a thermocline, but when you dive through one and the water goes from comfortably warm to shivering cold within a matter of feet, the phenomenon becomes truly unforgettable. Similarly, when diving on a coral reef or a kelp bed, one becomes acutely aware of the influence of ocean currents and waves. When going to the beach day after day to study sand movement, one becomes attentive to the truly dynamic nature of the shoreline. And when one encounters upwelling in the ocean on scuba, it is never more real or more interesting.

Upwelling in the ocean occurs when cool, nutrient-rich water wells upward from below into shallower water. Areas where upwelling occurs on a regular basis are some of the most productive in the world and can host a great abundance of marine life. Upwelling is responsible for the famous anchovy fishery off Peru and the historically rich marine life along California's coast. On a survey dive in the Florida Keys, my colleagues and I once came across a probable upwelling event. We first noticed a distinct shimmering or blurriness in the water that denotes an interface between two layers in the sea (a fascinating phenomenon called schlieren). As we dove through the schlieren, skirting a clifflike coral reef, the water became significantly cooler—cooler than during our previous dives—and there was a notable change in visibility: all signs of an upwelling event (more in chapter 6).

Days spent on or under the sea are well worth the challenges of getting funding and planning as well as facing the obstacles that must often be overcome. And while we usually come up with new questions in the field, hopefully we find some answers as well.

THE ANSWERS

When it comes to a question or problem, science or otherwise, most people want a relatively quick, concrete answer or definitive conclusion—a smoking gun in some cases. This is rarely how science works; instead, it tends to be a slow, stepwise progression toward knowledge and understanding. As data accumulate and are interpreted, our comprehension grows and theories are refined. Sometimes as science progresses, old hypotheses are thrown out and entirely new ideas pursued. For scientists this evolution is exciting yet at times frustrating, but it is the accepted norm, with uncertainty being a part of the process. For the public, however, the slow gains of research and uncertainty can be less easy to accept and breed dissatisfaction. Red-tide research in southwest Florida provides an illustrative example.

In southwest Florida, red tide, an episodic bloom or abundance of the alga *Karenia brevis*, has become a serious concern for people living, working, or visiting at the coast. Red tides can result in smelly and massive fish kills; the deaths of manatees, dolphins, and other marine organisms; and the closure of commercial shellfish beds. Onshore winds may also bring the toxins produced by the red-tide organism shoreward, causing beachgoers respiratory discomfort or distress. The public, media, and politicians in southwest Florida want answers. They want to know what causes red tide, if humans are making it worse, and how to stop it—seemingly simple, reasonable, and straightforward questions.

With funding made possible partly because of the public and political outcry, researchers have made real progress in the study of red tide, but the cut-and-dried answers so many people seek remain elusive. We do know that red tide has been occurring for hundreds of years and is a natural phenomenon. We have learned a tremendous amount about the alga *Karenia brevis*: its biology, the toxins it produces, and the environment in which it breeds and blooms. For instance, red-tide blooms typically begin offshore and move toward the coast at the surface or in bottom currents. The red-tide alga is a surprisingly versatile and complex organism with the ability to adapt as its environment changes. But we still don't know what triggers a bloom, what keeps it going, or why it ends. To grow and reproduce, *Karenia brevis* seems capable of using just about any nutrient source available—decomposing fish, nutrients from the atmosphere, from the deep ocean, from the coastal zone, or in flow off the land. Is pollution in the form of excess nutrients in river outflow and land runoff causing red tides to intensify and occur more frequently? In some places, nutrients flowing off the land and in rivers have been directly linked to algal blooms and the low oxygen conditions they can create—the most famous example being the huge dead zone that forms each summer off the mouth of the Mississippi River in the northern Gulf of

Mexico. The data available in southwest Florida, however, as of the writing of this book do not show a definitive link between red tide and land-derived pollution. It is also possible that overfishing has reduced the population of organisms that graze or filter algae from the sea and that may have once helped to keep blooms in check.

There is significant pressure on scientists in southwest Florida to come up with not only an answer, but the "right" answer. Public health scientist Barb Kirkpatrick of Mote Marine Laboratory emphasizes the need for scientific rigor and discipline in red-tide research, despite the pressure for quick answers. She reminds people that hastily obtained results may satisfy critics, but in the long run could prove inaccurate and lead to costly, unsatisfying, or unintended consequences. Scientists in the area have been strongly criticized because they refuse to interpret the available data as some would like and blame pollution for the problem. My colleagues and I have been quick to emphasize that pollution and excess nutrients in our coastal waters are harmful and should be reduced independent of the red-tide problem. As for controlling red tide, researchers are investigating methods used in Asia, but there are important questions to be answered regarding the impacts of these techniques on other marine life and the environment, and their cost and effectiveness.

As is the case with many, if not most, scientific endeavors, the answers about red tide will come, but they will only come over time as funding and research allow. Hopefully, our understanding will soon evolve to a point where we can reduce and mitigate the impacts of red tide. An interesting side note and a wonderful example of how science works: while studying the toxins produced by red-tide algae in southwest Florida, scientists discovered a chemical that is now being developed as a drug for the treatment of cystic fibrosis.

When it comes to answers, one of the harder lessons to learn in science is that a negative result is just as valuable as a positive one. This premise is not always easy to accept nor is it especially satisfying, particularly for students conducting their first independent research project. And of course, it is pretty difficult to publish a paper when the result of your research is a faulty hypothesis. But it is just as important to learn how nature doesn't work as it is to learn how it does. Certainly getting a negative result is always preferable to no result at all. I'd rather have an analysis come back different than expected than to have lost all of my sampling gear.

Along with a definitive conclusion, people also tend to want to categorize answers hierarchically. In science and the ocean, we often find that problems have multiple and interacting causes. It is not always possible to identify one cause as more significant than another. This can make finding solutions to a problem more difficult and the answer somehow less satisfying.

A good example illustrating this issue comes from the world of coral reef research. In recent years coral reefs across the globe have shown significant signs of change or, in anthropomorphic terms, ill health. Research has documented significant declines in live coral coverage and fish, episodic events of coral bleaching and mortality, overgrowth by algae on once coral-dominated reefs, and increased incidence of disease in corals and sponge. Overfishing, global warming, nutrient and sediment pollution, and destructive fishing practices are just some of the problems afflicting the world's coral reefs. The influences on coral reefs can also interact to make matters worse. For example, warming seawater temperatures may make corals more susceptible to disease and less resistant to or able to recover from storms. Overfishing can remove important herbivores on a reef and set the stage for fast-growing algae, fueled by pollution, to overgrow and smother corals. Yet people want to know, and spend precious time arguing over, what is the single worst thing causing the decline of coral reefs.

An important note here: sometimes it is not more answers that we need, but better use of the information we already have. We do in fact have a growing understanding of how human activities negatively impact the ocean, but we have only just begun to use this information to do what is necessary to better protect and preserve the sea, such as creating and implementing more effective ocean policies, providing sufficient investment, improved education, and better management strategies (chapter 8).

There remain many unanswered questions about the ocean. How will the ocean respond as the climate changes? How can we better manage human uses of the sea while preserving its health and abundance for future generations? Are coral reefs really dying, or are they resilient enough to adapt as the ocean changes over time? What ecological changes are taking place in the ocean due to human influence? Why do marine mammal strandings occur? What causes coral disease, and can we prevent it? Where do tuna and some of the sea's other majestic and commercially important species spawn or breed? What creatures or geological features remain hidden in the unexplored reaches of the sea? Without doubt, there are also many questions that we don't even know yet to ask.

Doing science is not a simple, boring, or orderly process that lacks creativity or adventure—far from it when studying the ocean. As scientists we work in a process-oriented approach, with a question or problem identified and a hypothesis developed to guide our search for a reproducible solution or an answer that stands up to scrutiny by our peers. Rarely does what we find in nature, however, exactly match what we propose while in the safety and isolation of our offices. Fieldwork is needed not just as a means to collect data or deploy equipment, but also for all that comes with seeing and being immersed in nature firsthand. My colleagues and I feel fortunate to have spent time doing science in, on, and under the sea. Our experiences in the field have taught us a tremendous amount about the ocean and about doing science, and as the following chapters reveal, they also make for some excellent sea stories.

2   *Tales of Wonder*

IN THE OCEAN THERE IS WONDER AND
mystery to be found. For my colleagues and me,
the sea's marvels pique our scientific curiosity and
make us appreciative of evolutionary complexities. They can
also provide an inspiring sense of discovery and at times just make
us stop and take note of the ocean's natural beauty and its seemingly
endless parade of strange and mysterious life-forms.

For many people, just the sight of a dolphin, a whale, or even
a shark can create a treasured lifelong memory. These, along with
furry sea otters and playful sea lions, are the ocean's charismatic creatures. Somehow, a scaly bug-eyed fish or gelatinous jelly just doesn't
seem to affect people the same way. For scientists that regularly go
into the field, encounters with the ocean's animals, charismatic or
not, are often memorable and sometimes humbling. Having evolved
to a life on land, our comings and goings in the ocean are nothing
short of awkward and laborious. When underwater, we must bring
our own supply of air and protection from the cold, the wet, and
the increasing pressure with depth. We must use fins or a motor to
propel ourselves with any sort of efficiency. Such inadequacies for a
life undersea are never more apparent than when we are confronted
by the agility and grace of the ocean's own animals. When a sea lion
swims effortlessly by, somersaults, hovers, and then blows bubbles in

your face, it is a memorable moment, and one that highlights your
own ineptness in the sea. When a dolphin or whale flicks its tail ever
so slightly and is propelled powerfully forward, we are left amazed
yet far behind, feebly kicking our feet or trying to start an engine.
Such encounters make us more conscious and appreciative of how
marine organisms have evolved so well to a life within the sea.

Personally, even a small squid can make me feel completely
inadequate in the ocean and in awe of nature's evolutionary ac-
complishments. Reef squid are typically less than a foot (0.3 m)
long and have an elongated baglike body, a small rounded head
with large eyes (one of the most advanced in the animal kingdom),
and ten tentacles in front—actually eight arms and two tentacles.
They also have thin, translucent fins that flutter along their sides,
which provide stability. By pumping water through their body,
squid can hover effortlessly or become instantly jet-propelled. They
are also unequaled masters of camouflage and can change color
in the blink of an eye. I have witnessed a squid go instantly from
wholly translucent to polka-dot to pulsating with waves of red
running through its body. Such versatile and complex traits make
our own two-legged gait and monochromatic skin seem downright
primitive.

Sometimes reef squid will approach if one stays still and quiet
while underwater. They often travel in small groups that seem
entrenched in an ongoing game of follow the leader. A lead squid
may hover or move ever so slowly forward as the other squid literally
line up behind and do the same. The leader may then wave a tentacle
or two or change color as the others follow suit. They never get
close enough to touch or hang around for long, but the company
of the reef squid is a wondrous, humbling thing to behold, if even
for just a few short minutes.

Other encounters with marine life have left me similarly intrigued and educated about the undersea world. Once while taking
underwater photographs, I braced myself on what I thought was

a rock only to soon discover a very surprised and sleepy sea turtle. Though sea turtles must regularly go to the surface to breathe air, some species can sleep underwater for up to five hours. The first time I encountered marine iguanas was another educational, if not slimy, experience.

Marine iguanas are found only in the Galápagos Islands. They are fascinatingly primitive-looking creatures, with leathery, wrinkled black skin, a small head and eyes, a pointed ridge down their back, and a body that tapers to a relatively long tail. They are usually found piled one on top of another basking in the sun along the islands' rocky shores. Marine iguanas are named for their unusual ability to dive into the sea to feed on algae. As swimmers they are strange to behold, using a sideways swishing of their tails for propulsion. Their gastronomic ventures into the ocean, however, result in a bit of a dietary dilemma—too much salt. Thus, marine iguanas have evolved an interesting means of eliminating excess salt. They sneeze it out of their noses. If you get too close to or startle marine iguanas, they tend to sneeze more, and one must then be wary of flying lizard snot—I learned this in the field as well.

While the sea's creatures can amaze us, sometimes humans can equally inspire disbelief. At a marine laboratory in the Bahamas, my colleagues and I were once called upon to help when a pilot whale stranded on a nearby island. My friend and coworker Wendy Keith-Hardy, her stepson who happened to be visiting, and I responded. Arriving at the island, we found a relatively small pilot whale stuck on the beach. Whales and dolphins strand for a variety of poorly understood reasons, including sickness or injury, chasing fish or others in a pod into shallow water, or errors in navigation. No matter what the cause, to survive they must be returned to the water as quickly as possible. Large marine mammals dehydrate rapidly, and their body structure can't support their massive weight on land.

The whale was what we expected; the line tied from its tail to a rock on shore was not. Our first job was to convince a local gentleman to untie the rope before it caused further damage. His plan, it seems, was to keep the whale in a pen that he would build adjacent to his bar/restaurant to attract customers. We quickly persuaded him that no one would come to see what would soon be a very dead whale. Did he really think it would survive long enough for him to build a pen? And then?

We untied the line and as the tide rose were able to gently push the pilot whale off the beach. Once in shallow water, it seemed disoriented and had trouble swimming. A short time later the small whale was moving about more steadily, but still seemed confused. We tried to guide it toward a channel that led to deeper water, but the whale repeatedly swam toward areas that would lead to another stranding. We then tried a new strategy. Wendy steered the boat on one side of the whale, while I got in the water and swam with snorkeling gear on its other side. I will never forget being eye-to-eye with that pilot whale, nor the power of its tail stroke. I had to take great care not to get within striking distance of its fluke. As I swam next to the whale, I hoped that it somehow knew we were trying to help. It didn't go well at first, and the whale kept swimming in the wrong direction—some cattle-herding experience would have come in handy on this one. Thankfully, success came hours later when, after reaching deeper water, with one powerful thrust of its tail, the whale dove down and disappeared from sight. As we made the long boat ride back to the marine laboratory, we were all exhausted, yet rejuvenated by the experience and its outcome.

A VOYAGE TO REMEMBER

My friend and colleague Captain Phil Sacks tells of his own amazing whale encounter. Unfortunately, this whale was dead, but the experience is nonetheless one that he remembers well during an eventful voyage and amid a career chock-full of at-sea adventures. In 1986 Phil was working aboard a 93-foot motor-sailing yacht in Greece. The yacht was a U.S.-flagged ship, and after America's military bombed Libya, the crew was directed to return to the United States as quickly as possible, hopefully in time for a planned tall-ship event in New York to celebrate the Statue of Liberty's birthday and restoration.

Just as they began shopping to provision the yacht for the voyage from Greece to New York, another international incident occurred—a major radiation leak from the nuclear reactor at Chernobyl. With winds blowing from the north, Greece was downwind of the leak, and people mobbed the markets in a panic, concerned about the safety of the food supply. Upon seeing the amount of food being bought by the ship's crew—for seven people for six weeks at sea—the locals grew angry, assuming they were hoarding. The crew tried to explain, but their efforts were hampered by their poor command of the Greek language. They left Greece quickly, sailed to Majorca, Spain, then to Gibraltar and across the Atlantic to the Azores, and on to Bermuda, heading for New York.

For Phil the passage from the Azores to Bermuda was particularly memorable. They were one day out of the Azores and awoke to a calm sea with "a glassy ocean surface and just a slight greasy swell running." With such a calm sea, the light plays tricks, and it can be hard to distinguish the horizon. On the other hand, floating objects are easier to see, and in the monotony of an ocean passage, sometimes just investigating a buoy can be an interesting distraction. That morning the lookout noticed something floating at the surface off the port bow, maybe a quarter mile away. They altered course slightly and approached the object. "It was a dead whale, bloated with air. But what really surprised us was that there were two large sharks swimming around the whale," recounts Phil. The sharks weren't just swimming around the carcass; they were actually sliding right up onto its back, almost to a point of being completely out of the water. It was a dead sperm whale, and while Phil can't recall what type of sharks they were, he is sure they were large, deep-ocean predators about 10 to 12 feet (3 to 4 m) long.

At their previous port stop in the Azores, Phil had been in a bar in the old whaling port of Horta where an abundance of whalebones and teeth had been on display. Inspired, he decided to try to collect a few teeth from the whale carcass. Given the large sharks swimming nearby, Phil decided it probably wasn't a good idea to launch their small boat for the job—it would be safer to attempt a tooth extraction from aboard the yacht itself. The sharks, however, did not want to share, no matter what boat they used. As the crew pulled alongside the whale and rigged up a line to lift the whale's head closer to their reach, the large predators grew more aggressive, a little too aggressive, and in the end, the sharks won. Phil relinquished the teeth to a burial at sea when the whale eventually sank into the ocean's depths. As they drifted away, he still remembers being utterly fascinated by the sharks' behavior. In case you too ever happen upon a dead whale, it is now illegal to extract the teeth, or to trade or buy any part of a marine mammal.

As the yacht continued toward Bermuda, the swells disappeared completely, leaving the ocean as smooth as a sheet of glass. There was also a full moon, and for three full nights Phil recalls another amazing sight at the sea surface: "It was filled with hundreds of gelatinous animals, spiral forms, maybe over 12 feet [4 m] long." He never saw the organisms during the day, but each night the sea was once again replete with the long, transparent creatures. There was no net aboard to collect a specimen, but after the voyage Phil described the organisms to biologists. He now thinks it may have been a massive aggregation of simple, barrel-shaped creatures called salps. Relatively recent observations from submersibles suggest another likely candidate—siphonophores. These are long, delicate, gelatinous creatures that can form chains up to 100 feet (30 m) long. To this day, Phil remains amazed by the encounter: "I have probably spent something close to 3,000 nights at sea, but have never again experienced anything like it."

NIGHTTIME IN THE SEA

After dark the sea is indeed a very different place than during the day. Underwater, just the progression from day to night can be striking. As dusk falls, the shallow ocean begins to turn dark, and it becomes a time of shadows, when many of the sea's visual predators most actively seek their prey. As darkness more fully overtakes the ocean, nocturnal animals begin to emerge while their daytime counterparts seek refuge. In the full of night with heavy cloud cover or a new moon, even the shallow sea turns inky black. However, on moonlit nights the upper reaches of the ocean can remain amazingly bright. On one late-night boat ride across a shallow bank in the Bahamas, the moonlight was so bright, the water so clear, and the underlying sand so white that we could see stingrays and fish swimming below as we raced by. The nighttime ocean is particularly dramatic on a coral reef.

At night many creatures on a coral reef, such as octopus, moray eels, and the light-shy squirrelfish, swim about more freely. And corals, most of which look like lifeless rocks during the day, become more active after sunset. Most corals are colonies of small animals called polyps, each of which is essentially a stomach surrounded by a ring of short tentacles. At night the coral polyps extend their tentacles into the water to feed on particles and small organisms drifting by. When a diver's light is held over a coral at night, it creates an astonishing sight. Zooplankton and small fish attracted to the light soon fall prey to the coral in what can only be called a tentacular feeding frenzy. At night under the glare of a light, colors on the coral reef also appear enhanced: the red of a sponge becomes rich and especially vibrant, the green of an octopus or moray eel turns luminescent, and the yellow of a seahorse seems startlingly bright.

Marine scientist and longtime diver Steve Gittings is currently the science coordinator for the National Marine Sanctuary Program. His choice for one of the ocean's most amazing sights is a nighttime wonder that can be seen only once a year at a very specific time—it is the mass spawning of corals. Mass coral spawning is a somewhat mysterious phenomenon that was first discovered on the Great Barrier Reef in 1982. In the Gulf of Mexico's Flower Garden Banks National Marine Sanctuary, it occurs eight days after a full moon in August, at around nine o'clock at night. Nearly all at once, a huge number of boulder or head corals release countless bead-sized bundles into the water. Steve says, "The crystal-clear water is turned into a soup of ascending spheres that envelopes awestruck divers, who then rush in to record the event with all sorts of cameras and data collectors." Each bundle released contains both coral eggs and sperm, but somehow the sperm doesn't fertilize the eggs from the same parent. Once the bundles have floated to the surface, they break apart, and the sperm can then find an egg from another parent coral, "putting into motion the life of a potential new coral colony."

While mass spawning is incredible to observe, Steve also finds the underlying reproductive process fascinating. Throughout the world, many reef corals reproduce only once a year, and while this seems simple enough, it is complicated because absolute synchronization among individuals is essential. For the corals to successfully cross-fertilize, gametes (sperm and eggs) from different individuals must be released at nearly the same time; otherwise, viable offspring will not be produced. For Steve it is an amazing coral choreography, one that takes place without eye contact, words, or even an enticing mating dance.

Successful mass spawning of corals may also be an indicator of their health and potential to remain viable. With so many of the world's coral reefs showing evidence of degradation, Steve believes that continued mass spawning on the Flower Garden Banks is a good sign. He also strongly believes that ensuring good water quality over a reef is critical to nature's own approach to coral reef management and proliferation—mass spawning.

The Flower Garden Banks National Marine Sanctuary was established to protect and allow for the study of the region's coral reefs and other undersea life. Because of its location, some 110 miles (177 km) south of the Texas and Louisiana coasts, the sanctuary has become a popular diving and research destination. Yet only a lucky few scientists or adventurous divers get to witness the wonders of mass spawning at the Flower Garden Banks, on one very special night of the year.

AN UNDERSEA GLOW

Nighttime also provides the opportunity to observe one of the most spectacular and fascinating displays in the ocean—bioluminescence, or the production of light through biological processes. Researchers have discovered that a surprising number of the sea's creatures have the capability to produce light. It occurs in a variety of forms; the most common is a tiny twinkling created by small planktonic creatures called dinoflagellates, which typically produce light in response to water motion. A producer on a television show about the Bermuda Triangle once asked me about a mysterious glowing sphere reportedly floating near a boat just before it was lost. She was surely hoping for a dramatic response, something along the lines of the supernatural, aliens, or an unexplained mystery. My answer— jellyfish, which create another common form of bioluminescence. They often produce spherical bursts of bright blue-green light. Though it is not fully understood, scientists believe that organisms in the ocean use bioluminescence for several purposes, including as a decoy to escape, to hide from or startle predators, as a form of communication, and as a means to lure in prey.

Deep-sea biologist Edie Widder, cofounder of the Ocean Research and Conservation Association, was originally planning a career in molecular or neurobiology. One wondrous dive in the deep sea, however, changed her career path and led to her becoming one of the world's leading experts on bioluminescence. It was the early 1980s, and she was doing her first dive into the deep ocean in a tethered deep-diving suit, which she likens to being the yo-yo on the end of a very long string. Edie expected to see some bioluminescence in the water during the dive, but when she turned off the suit's outer lights, the display around her was staggering. The sheer number of light-producing creatures was far beyond anything she had ever imagined. As she stared at the bright flashes occurring throughout the water column, she pondered the role of bioluminescence in the sea. Edie vividly remembers thinking that if so many marine organisms can create light, it must play an important, yet unrecognized, role in the ocean. It was a life-changing moment, and she has since spent her career studying and quantifying bioluminescence in the sea. She is now using this expertise to help conserve coastal areas. Her story about using fake bioluminescence to lure in a new type of squid is in chapter 5.

Bioluminescence is also a great crowd pleaser. Aboard SEA ships, a large net is regularly towed deep in the sea at night to collect organisms for teaching and research. The midwater or twilight region of the ocean, between depths of about 600 and 3,000 feet (200 to 900 m), is typically targeted. Many of the sea's animals migrate upward into shallower water at night; hence nighttime tows enable easier collection of deeper-water species. The highlight of such tows is unquestionably the display of bioluminescence during the retrieval process. As the net is pulled in, a ghostly green glow appears far off the ship's stern. As the net gets closer to the surface, the glow intensifies, and quick bursts or flashes of green light occur within. This incredible light show often continues until the net is under the glare of the ship's lights.

Once the net is onboard, there is the fun of seeing what strange creatures have been ensnared. Students wash, pick, and sieve through all that is caught, counting and identifying as much as possible. Some of the organisms are easy to identify, such as the bright red shrimp that live in the midwater region of the sea. A toothy viper- or hatchetfish is always an interesting find—luckily they are just a few inches long. Other organisms are less easily identified, having been smashed by the effects of the net tow—much like a washing machine's spin cycle—and the result is typically copious amounts of slime, fondly referred to as sea snot.

On one cruise a surprising example of blue bioluminescence sparked a late-night adventure in marine biology. While helping students pick through the results of a nighttime net tow, I observed a small flash of brilliant azure blue in the bucket. Startled, and curious as usual, I had to know what had created the flicker of light. We watched and waited for it to happen again, at the ready to collect the culprit. With the next flash of blue, we tried using a small net, a spoon, and even an eyedropper to scoop up the responsible creature. I honestly don't remember how we finally caught it, but the challenge definitely tested our patience late into the night. Our efforts were finally rewarded when we got the animal into a petri dish and under the microscope. It was small, about half an inch (a centimeter or so) long, oval-shaped, and appeared as if bejeweled with colorful gems. Each of the animal's multiple segments was a different and vibrant color. Unfortunately, the spectacular tint soon vanished, and the tiny animal became entirely transparent. The students and I spent hours searching through the ship's library of books to identify the beautiful creature. We eventually found it, a copepod aptly named *Sapphrina*. Although the organism was not new to science, for us it provided a marvelous evening of wonder and discovery.

A SENSE OF DISCOVERY

For oceangoing scientists, wonder often goes hand in hand with discovery. Relatively few scientists discover a new species or are the first to explore a particular habitat. Yet for most scientists, myself included, we feel an inspiring sense of discovery when seeing things for the first time, whether they are new to science or not. Caroline

Rogers, a marine ecologist for the U.S. Geological Survey in the U.S. Virgin Islands, has spent years studying Caribbean coral reefs. Of the first time she dove on a coral reef in the Pacific, she says, "It was nothing less than a spiritual experience." Seeing the unmatched diversity of color and species of coral in the Pacific as compared to the Caribbean provided her with an incredible sense of discovery.

Chuck Messing, a marine invertebrate zoologist at Nova Southeastern University's Oceanographic Center, studies crinoids, the ocean's flowerlike sea lilies and feather stars. Whether on scuba or in a deep-sea submersible, he is always on the hunt for a crinoid, or any invertebrate, that he has never seen or studied before. On a trip to Australia's Great Barrier Reef, one of his most memorable scuba dives was in an open sandy area, because there he found an unusual species of crinoid. Chuck notes, "Who goes to the Great Barrier Reef to dive in sand?"

Bob Ginsburg vividly remembers the first time he landed on the west side of Andros, an island in the Bahamas. For years he had been studying thick sequences of ancient rocks on land made of fine calcium carbonate, which exhibited strange features, such as filled-in cracks and intraclasts (chunks of a lower deposit found "floating" in the rocks above). At Andros, Bob found that the island's extensive tidal flats were covered with mud cracks and loose chips of mud; these were the precursors to the features he had been puzzling over. He was face-to-face with the modern equivalent of his ancient rocks. Given the great expanse of the corresponding rock formations, Bob also realized that tidal environments similar to those on Andros were much more extensive in the past than they are today. He published a series of research articles based on those early days at Andros and still remembers the feeling of awe and discovery during that very first visit.

Shirley Pomponi, a deep-sea biologist at the Harbor Branch Oceanographic Institution, recalls the incredible sense of discovery she felt when seeing deep-sea sponges from a submersible for the

first time. She was also amazed to find that the sponges looked just
like the drawings done of them—100 years ago. The artists had
never actually seen what she was seeing—the deep-sea sponges
in their natural environment. Their beautifully colored drawings
were based only on dredged samples, which must have been heavily
damaged and filled with mud. Yet incredibly, the sponges Shirley
saw in the deep sea looked just like the old illustrations.

OOOO-IDS

Things in the ocean that are cute, big, scary, or glowing are a sure
hit with just about any audience, but it is a bit more difficult to
convince others, especially students, that something as simple
as sand can be equally as delightful. During my tenure at SEA, I
conducted experiments on the dynamics of carbonate and quartz
sand grains, testing the flow strength that would initiate transport
in a runwaylike flume at the Massachusetts Institute of Technology.
To collect sediment samples for the experiments, I enlisted the help
of my students on an SEA cruise. I was particularly enthused about
a certain type of sand from the Bahamas. The students, however,
did not share my zeal, often gracing me with the classic rolling of
the eyes when I spoke about the *magic of ooids*.

Ooids are small, white, beadlike grains of calcium carbonate
that form in only a few places in the world today, but were more
prevalent in the geological past. Underlying Miami, for instance,
is a rock formation appropriately named the Miami oolite—a
rock made of ooids. One of the largest oil fields in the world today
is a reservoir of ancient limestone composed of—ooids. Over
the years, geologists have debated whether ooids form strictly by
chemical precipitation from water supersaturated with calcium
carbonate or through biological processes facilitated by bacteria.
Either way, these very round, very white beadlike grains form as
multiple layers of calcium carbonate crystals accumulate around
a nucleus. Conceptually, ooids are simply interesting, but when

one jumps out of a boat and sinks knee-deep in billions of them it is truly special—really!

The SEA ship arrived at Joulter's Cay in the Bahamas, and with my skeptical students at the ready, we launched a small inflatable boat on a mission to collect ooids. The students continued to shake their heads and roll their eyes over my excitement as well as the large collection of plastic tubs and bags that I had brought along for storing samples. We motored to the shallow water just off the beach, and I enthusiastically jumped out of the boat, sinking into billions of ooids. The students handed me several bags and containers, which I began filling with the beadlike grains. Several students then decided that they too wanted to experience ooids firsthand. Jumping out of the boat, they sank knee-deep, and their expressions were nothing short of glee. Soon they also wanted bags for collecting—it turns out that ooids are pretty cool, even for the nonscientist.

AN ARMY OF SEA URCHINS

In 1997 marine biologist Bill Sharp of the Florida Fish and Wildlife Research Institute was investigating an unprecedented bloom of sea urchins (*Lytechinus variegatus*) in Florida Bay. He asked me to examine the geological effects of the event. I went into the field expecting to see just a few more of the prickly echinoderms than usual. What I found was nothing short of astonishing. There had to be hundreds, so many sea urchins that they were literally piled up, one on top of another. Not only was their sheer number surprising, but so was the impact they were having on the seafloor. It was as if an army of out-of-control lawn mowers had marched through the seagrass. In front of the sea urchins stood a meadow of lush, long-bladed green seagrass, while in their wake all that was left was bare sand and mud. From a geological perspective, the grazing by sea urchins had left the sediment more vulnerable to erosion and resuspension, but why the population bloom occurred and how often it happens remains a mystery. It was a sight I will never forget.

THE WEST WALL

Linda Glover, an oceanographer and policy expert who spent many years working for the Oceanographer of the Navy, tells of her own at-sea wonder. At the time she was using satellite images and ship observations to make weekly estimates of the position of the Gulf Stream and its eddies for the National Weather Service. Linda decided to see just how accurate their position estimates were during a recreational sailing trip from Norfolk, Virginia, to Bermuda. With a fax of their latest analysis in hand, as they were sailing east and approaching the predicted position of the Gulf Stream, she began using a bucket over the side and a thermometer to measure ocean temperatures. It was a calm, sunny afternoon, and she expected to detect an abrupt rise in temperature as they hit the "West Wall" of the current. "Looking ahead of us, we suddenly saw the wall," she recalls. Stretching along the horizon for as far as she could see in both directions was a distinct change in water color and a literal step up in the sea surface. The warm, lower-density water in the Gulf Stream had actually created a small wall on the current's western edge. Linda estimates it was about a foot (0.3 m) or so in height. Happily for her, the actual position of the West Wall also coincided fairly well with their predicted location based on the satellite data. Just the right conditions were needed for Linda to actually see the Gulf Stream so dramatically; more often than not it is visually undetectable and sometimes can create truly nasty sea conditions for boaters.

A BALLET ABOARD SHIP

Mel Briscoe, a longtime oceanographer with the Office of Naval Research, an expert on interagency ocean policy, and a dive instructor, has a different, though no less interesting, perspective on an ocean wonder. To set the stage, he explains that in the 1960s there was a new vision in physical oceanography—to measure deep-sea currents over a long period of time with instruments and data loggers attached to ocean moorings. Moorings are essentially a long cable with oceanographic instruments attached at points along its length. Each can be several miles long, with a large buoy at the surface, and is anchored by a heavy weight at the bottom. Prior to that time, oceanographers had used only primitive mechanical equipment for current measurements in shallow coastal waters for just a few days.

Mel likens the safe launching of a mooring in the deep sea to a beautifully choreographed, but dangerous, ballet. The stage was the fantail of a ship, an open deck where large, heavy equipment could be lowered into or lifted from the sea. It was a venue that could be rolling in the seas or awash with water as waves broke over the side. And the mooring anchors weighed tons, literally, while the buoys were large and cumbersome to move. It was precarious work and no place for the lighthearted. The players were, as Mel puts it, "strong men in rubber boots with knifes on their hips, directed by the chief scientist, and stage-managed by the ship's bosun."

For Mel the deployment of a mooring by the ship's crew was a true wonder. A large buoy would first be lifted by a crane, swung over the side, and lowered into the water with the upper part of the mooring line already attached. The crew used lines that went through a cleat to the buoy to safely keep it from swinging uncontrolled and becoming a deadly battering ram on deck. While one person handled the line, a second watched to ensure that no one was standing in a loop, or bight, that could ensnare a foot and, in an instant, drag that person over the side. Safety railings or lines at the deck's edge were forfeited to avoid interfering with the equipment going over the side. Instead, each deckhand working near the edge had a person standing behind them hanging on to their belt, just in case.

The mooring line was spooled out from a large winch as the ship moved slowly toward a designated research site. At specific spots along the cable, as determined by the research plan, a current meter was attached. Mel recounts, "It took five deckhands to dance this part of the ballet. Four to attach the instrument and one observer making sure no parts were missing or faulty." He remains amazed that only a few moments were needed for the ship's crew to attach a current meter and then continue, as the cable was rapidly spooled out. It took hours for a mooring to be deployed in its entirety; each could be 16,400 feet (5,000 m) long and have twenty current meters attached. Interspersed along the cable were also large flotation devices—heavy glass balls in plastic shells—to balance the weight of the mooring line underwater. Throughout the process the actors in this ballet would change, substitutes made as someone went off the deck for a rest or to refuel.

Just before the final act, the anchor would be attached. By this time the mooring was stretched out behind the ship for miles, floating into the distance with the line, current meters, and glass balls all being dragged along. At the designated position for the mooring, the anchor would be pushed over the side and instantly drop to the bottom, pulling the cable, instruments, and buoys along with it.

With a well-performed ballet, the mooring would be in the right location and at the right depth, and all the instruments would be in working order. Given the high cost of ship time, the crew would work night and day deploying moorings on each cruise, sometimes up to twelve. A year later the ship would return to recover each mooring, and the scientists would download their data from the instruments. The recovery operation required another ballet, and this one could be even more dangerous. Mel fondly remembers his time at sea deploying moorings: "It was all quite beautiful to watch and exhilarating to be a part of."

These days similar long-term mooring deployments are still done, but now instruments can be programmed to transfer their data to a buoy at the surface, which can then transmit the data via satellite to laboratories on shore and across the globe via the Internet.

WATCHING WHALE SHARKS

Two last, equally wondrous and thought-provoking tales about the marvels of the sea—in this case very large ones, namely, whale sharks. The first comes from Steve Gittings and an experience in Belize several years ago while he was working with colleagues study-ing spawning aggregations of groupers. They were diving at just about sundown, and the ocean was starting to turn dark. Steve could see schools of large fish gathering below—fish that usually swim alone on a reef. Soon there were hundreds, maybe even thousands, of these fish, slowly rising toward the surface, in one enormous and growing school. Steve was surprised as all around him "they swam in slow circles at first, gradually increasing their speed and forming tighter and tighter spirals, creating a living tornado." He was a bit disconcerted as the fish got closer and closer, especially the four-foot-long cubera snappers with exceptionally large, sharp teeth protruding from their mouths, clearly oblivious to everything but spawning. They were constantly changing color, becoming dark and then light, until the apex of activity when they released eggs and sperm into the water. The visibility for the divers instantly became negligible, and then as Steve watched in "shock and awe," out of the blue lumbered a giant whale shark, its huge mouth opened wide as it swam slowly by, dining on the surrounding soup of caviar. Then another whale shark swooped through, only to be followed by more of the monstrous planktivores. During his twenty years of diving, Steve had seen a few solitary whale sharks, but he had never encountered anything like this. It was truly one of the highlights of his career in marine science—so far.

The experience also made Steve think about a recent report suggesting that a high percentage, possibly as great as 90 percent, of the large fish in the ocean have been fished out. Was he witnessing an event that was once fairly common in the sea? If overfishing continues, would his kids know about such miraculous spawning events and whale sharks only as stories told about times gone by?

The second, no less awe-inspiring, whale shark tale comes from marine biologist and shark expert Bob Hueter of Mote Marine Laboratory. For over ten years, every spring Bob and his colleagues had been going to Mexico to study sharks. The shallow lagoons of the area are nursery grounds for blacktips. With long hours in the field and by working with local researchers and fishermen, they were able to learn a great deal about the sharks' life cycle, behavior, and movements. One day, about four years ago, while sitting in a boat, Bob was chatting with a local fisherman who had become a trusted and helpful partner. Out of the blue the fisherman said something to the effect of "Bob, you do know that each year after you leave there are whale sharks and manta rays all over the place?" Bob was stunned and couldn't believe that no one had ever mentioned it. He had always wanted to study whale sharks, but they tend to be hard to find.

Four years later, over the course of essentially three summers, Bob and his team of researchers had tagged some 550 individual whale sharks in an area now believed to host the largest aggregation of whale sharks in the world. From May to September of each year, perhaps 1,500 whale sharks congregate off Isla Holbox, Mexico, to feed on an abundance of plankton produced by upwelling in the region. Bob recalls one field session in particular with the whale sharks. The enormous creatures were in an area of exceptionally clear water, and his team was in the midst of tagging, measuring, and collecting as much data as possible. In a moment of inspiration, Bob instructed the team to drop what they were doing and simply swim, watch, and appreciate the magnificent creatures all around them. He says it was unforgettable, "like having three 35-foot buses at your sides. Yet they are gentle, graceful animals unbothered by your presence." By just stopping and taking the time to appreciate and observe the whale sharks, Bob was also able to learn about them. He had previously thought that their eyes were located strictly on the sides of their head, which would create a puzzling problem—a huge blind spot at their front. How then, did these enormous animals avoid running into things as they swam slowly through the sea filtering out plankton? With closer observation that day, he realized that their eyes are in fact rolled slightly forward and to the side; they can see what lies directly ahead without a problem. There is great power in simple observations in the field, yet sometimes we forget how much there is to learn just by watching. Bob's story also illustrates the benefits that scientists can obtain from working with local people who spend much of their time in the field and are willing to share their knowledge.

~~~~~~~~~~~~~ From biological marvels to surprising physical and geological features, the sea is a place of wonder. For scientists, encounters at sea bring to life what we spend hours and hours reading about in a library or studying in an office or laboratory. Such occasions make us contemplate how organisms have evolved and adapted to a life in the ocean and about our own relationship to and impact on the sea. Experiencing the ocean's amazing wonders firsthand is one of the great and precious benefits of doing science at sea.

3 *The Unexpected*

SERENDIPITY AND ENCOUNTERING THE unexpected may be two of the most important, yet underappreciated, aspects of doing fieldwork in the sea. Unforeseen or chance encounters in the field often lead to surprising insights, dramatic shifts in our understanding, and moments of inspiration or enlightenment. This happens partly because when scientists go into the field, they have certain expectations about what they will find; these result from earlier experiences, academic training, or what has previously been published. Yet any good researcher knows that they must also remain open to new ideas or the possibility that their previous assumptions are wrong, because in the ocean we don't know what we don't know. The following stories come from several disciplines of marine science and provide just a few instances that showcase the important role that serendipity and the unexpected play in fieldwork and in science in general. The first is a terrific example from a marine biologist at the Rosenstiel School, in which a eureka moment underwater not only altered his ideas, but also profoundly influenced his career as a scientist.

"I JUST FOLLOWED THE FISH"

Mike Schmale is one of the world's leading experts on the tumor viruses that infect fish on coral reefs, in particular, damselfish. He began studying coral reef fish while investigating their behavior for his master's degree. As a graduate student, he spent lots of time underwater conducting reef fish surveys, and every so often would find bicolor damselfish with conspicuous, heavily pigmented spots and lumps. Intrigued by these deformations, one day he delivered several of the motley-looking creatures to George Hensley, a pathologist with a strong interest in the study of comparative diseases and a keen eye for diagnosing unusual tumors. An interesting diagnosis by Hensley could mean that Mike had found his next research project and a PhD dissertation topic. Fortunately for him, but unfortunately for the fish, the news wasn't good; they had malignant tumors and not long to live. The damselfish were suffering from the fish version of a tumor that can affect the peripheral nervous system of humans (neurofibromatosis).

Mike returned to the field in search of an explanation for the tumors. It was the beginning of several years spent largely underwater on reefs from Miami to Key West in pursuit of diseased damselfish. He spent hours and hours underwater counting fish. He even constructed an underwater fish-counting device—somewhat absurd, he says in retrospect—which enabled him to efficiently tally sick and healthy damselfish of different sizes while swimming about the reef. It was basically an underwater abacus made of stainless rods and rubber faucet washers attached to the back of an aluminum clipboard. He estimates having counted more than 40,000 damselfish.

At the beginning of his damselfish survey, Mike remembers going into the field with some preconceived notions about what he would find. At the time, the scientific literature suggested that chemical carcinogens in the water or sediments caused the vast majority of fish tumors. Naturally, he expected this to be the case for the damselfish tumors. "Many of my dives were spent wondering how in the world I would ever even begin to test such a hypothesis. I was swimming in clear water, six to eight miles offshore with

no signs of pollution in sight! This was the Florida Keys, not the
Niagara River or Boston Harbor, where tumors and toxicants were
easily matched."

Mike recalls feeling a bit hopeless as he swam back and forth
across one reef after another. Then he started to discern a pattern.
Damselfish are very territorial, preferring to stay in one spot on
the reef and defend it against neighboring damselfish as well as
other types of fish (or divers). This made it easy to count and map
the locations of sick fish relative to each other and to healthy fish.
Mike began to realize that the sick fish were bunched; they were
clustered in their distribution. If he found one sick fish, within a
few to tens of feet (a few meters) of it, there would almost always
be one or two more sick fish.

One day, while he was once again swimming over a reef, a new
hypothesis suddenly came to mind. The clustered distribution of the
diseased fish suggested an infectious agent, something like a virus,
rather than an environmental toxin—something that a fish could
pass to its nearest neighbor, perhaps while fighting over territories.
Mike's realization quickly changed the course of his research and his
professional career. He headed into the laboratory to try to transmit
the tumors from diseased to healthy fish under controlled conditions.
After several anxious months of waiting, the fish in his experiments
began to show early signs of the same malignant tumors he had
found in the wild. The tumors were indeed infectious.

Twenty years later and Mike is still working on damselfish
and tumors, though much of his work is now done in the laboratory—all based on a flash of insight while in the field. He just
followed the fish.

Mike is one of numerous scientists now striving to better identify and understand the links between the oceans and human health.
Researchers in this growing field are exploring the ocean in search of
marine-derived chemicals that can be synthesized in the laboratory
and used to fight human disease and illnesses. Sea squirts, colonial

bryozoans, and deep-sea sponges have already produced substances that are being used as effective drugs against certain types of cancers. Other marine organisms are providing substances that are used to battle inflammation and viruses and to detect toxins in medicine. And a recently derived painkiller comes from one of the ocean's deadliest creatures—the cone shell. Other researchers, like Mike Schmale, are using marine organisms as models to study human ailments. The Rosenstiel School raises and ships some 25,000 sea slugs (*Aplysia*) a year to facilities around the world so that scientists can use their extraordinarily large neurons and simple nervous systems in neurobiological research.

The ocean can also be detrimental to human health. Increasing incidence of harmful algal blooms, seafood poisoning, and waterborne illnesses puts people at risk, as do rising sea level, coastal storms, tsunamis, and increasing seawater temperatures that may cause a rise in and spread of diseases, such as malaria and cholera.

"A SPEARED FISH AND POTTERY SHARD"

From a more classical discipline in ocean science, marine geology, comes a wonderful story about an unexpected discovery in the field and the impact it had on science, a career, and the entire oil and gas industry. This tale comes from my good friend, colleague, mentor, and all around underwater adventurer, marine geologist Gene Shinn. Gene calls it serendipity. I call it being in perpetual field mode—constantly curious, continually driven to explore the world around us, and always questioning. In his case, it is also the willingness, or maybe the propensity, to go against the grain.

The story begins in the 1960s in what Gene calls the Golden Age of Research for marine geology, when major oil corporations were spending a large percentage of their profits on research. Much of the work focused on understanding how calcium carbonate sediments and coral reefs form in modern ocean environments, as an analog for those of the past. Ancient limestone (calcium carbonate) formations now buried thousands of feet beneath the surface are excellent reservoirs for oil and gas, so knowing how and where they are likely to form is advantageous to exploration, and potentially lucrative. At the time, within the scientific community and especially within the oil industry, the ruling dogma was that limestone forms only when sediments made of calcium carbonate are exposed to freshwater. This had a great influence on where the industry spent its millions in drilling for oil and gas.

Growing up in the Florida Keys, Gene spent years diving and boating around carbonate sediment environments, such as coral reefs, not to mention becoming a national spearfishing champion. With his extensive field experience, he was sent to the Persian Gulf to set up a small laboratory for Royal Dutch Shell. He was to investigate and map sedimentary deposits and coral reefs about the Qatar peninsula and participate as part of a team in Gulf-wide expeditions. Gene says of his time in the region, "There were few diving scientists, and this was truly fertile grounds for discovery."

Gene's work in the Persian Gulf got off to an interesting, seem-ingly unsuccessful, start during a large-scale expedition to collect rock and sediment samples. The geologists aboard the ship were having a hard time—literally. Their attempts to recover samples were continually spoiled by the hard rock bottom. Core tubes kept coming back squashed like accordions, and grab samplers repeatedly came up with only a handful of sediment. According to Gene, the well-trained geologists onboard all became convinced that there were few sediments being created or deposited in the modern Persian Gulf. They also assumed that the hard rock bottom was the result of freshwater cementation during exposure to the atmosphere and rain when sea level was substantially lower, about 15,000 years ago during the last glacial period. The entire Persian Gulf was presumed to have been dry at the time. Gene recounts thinking, "We were simply on an expensive wild-goose chase for sediment that did not exist."

During his off-hours, Gene, like many of us, couldn't stay

out of the water, and weekend camping trips with his friends and
family always included a bit of spearfishing for a fresh grouper
dinner. One day, while giving chase to a fish he had speared, Gene
noticed something unusual about a rock ledge: "Pottery, glass, nuts
and bolts, and other junk had become part of the rock." He was
puzzled by the discovery because the rock was 7 feet (2 m) below
the surface and the stuff cemented into the rock had clearly come
from the boats tied to the dock. It could not have formed 15,000
years ago.

Being in perpetual field mode, Gene collected a sample of the
rock "complete with pottery" and later, in the laboratory, prepared
a thin section to examine it more closely. Under the microscope he
saw a thick fringe of precipitated calcium carbonate that looked just
like what company reports and textbooks all showed as freshwater
cement. He sent a photograph of the sample to his supervisor in
the Netherlands, and a geochemist wrote back suggesting that
the cement used to make a nearby jetty had somehow created the
submarine rock.

Gene considered the idea but could find no evidence in the
field to support it. On another camping trip, he found a larger
piece of pottery protruding from hard rock; there were no jet-
ties in sight, and it was at least 10 feet (30 m) underwater. Using a
pick and hammer, he spent several hours removing a hunk of the
pottery-containing rock. Under the microscope, it proved to have
the same kind of cement as the first sample. He began to think
that the large expanses of hard rock they had found while diving
throughout the Persian Gulf might all be the product of the same
process—submarine cementation.

The proposition that limestone could form underwater was a
concept too intriguing for him to ignore, and if it was true, it went
squarely against the thinking of the time. Gene excitedly set out to
do some serious undersea digging. The bottom was exceptionally
hard, but with perseverance he kept at it and eventually discovered

several 4-inch- (5-cm)-thick layers of hard rock with loose sediment in between. Some of the layers were coated with amber-colored aragonite, a form of calcium carbonate known to precipitate from seawater. In the meantime, back at the main lab in the Netherlands, there was building resistance to the idea that Gene was finding rock forming underwater in the ocean. He was often reminded that he should be working on tidal-flat muds, dolomite, and gypsum instead of chasing his crazy idea.

When the first date from radiocarbon age analysis came back, Gene's case for underwater cementation received a huge boost. The rock was young, too young to have ever been exposed to the atmosphere or rainwater due to a drop in sea level. The evidence continued to mount and eventually led to an inescapable conclusion—limestone rock was indeed forming underwater in the Persian Gulf. Gene admits that while he knew he was on to something significant, he was not yet sufficiently trained as a geologist to know just how significant.

He continued his efforts to prove his theory and while diving with local pearl divers discovered the same hard rock bottom 60 feet (18 m) down. Gene then borrowed a few explosives from some visiting geophysicists to expose the layer of hard rock, which turned out to be a foot (30 cm) thick with pearl oyster shells cemented in it. Carbon dating proved that the rock was only 3,000 years old. It had to have formed under present-day marine conditions.

The discovery of submarine cementation set off a chain reaction in the geological community, and Gene found himself at the center of his first scientific controversy. A heated argument arose between geologists at Shell in the United States and those with the Royal Dutch Shell group. American Shell geologists at first denied the possibility of rock forming underwater in the sea; the company had invested millions of dollars into researching and refining the details of freshwater cementation.

Gene had to laboriously document all of his evidence for submarine cementation. He says it taught him a lot about how science works, "Science is very conservative especially if results do not fit a political agenda or the latest fad. New ideas are not quickly grasped." Unlike a theoretical model, however, Gene's observations in the field could not be disproved. For those in the scientific community who had already suspected that sediments could harden underwater, Gene's proof was welcome, and eventually the others, too, were convinced, with fieldwork being the key.

Today submarine cementation is taken for granted and explained in textbooks. Gene's discovery due to a "speared fish and pottery shard" had significant ramifications, including on where petroleum geologists look for oil and gas and on our understanding of how coral reefs harden over time. And when someone found a mysterious roadlike outcrop of rocks underwater in the Bahamas, who did they call to investigate the possible "road to Atlantis"? Gene Shinn, of course, who proved that it was simply rock that had previously formed on a beach (beachrock) and then became submerged as the sea level rose.

The experience in the Persian Gulf changed Gene's career and his view of how science is accomplished and accepted by one's peers. He is also a strong supporter of going to sea for science and like many of us fears that regulations, permitting, bureaucratic micromanagement, and a lack of funding are now making fieldwork next to impossible to conduct.

GOING OVER THE FALLS

From my marine geologist colleague Bob Halley of the U.S. Geological Survey comes another great story about encountering the unexpected underwater. Bob was working with colleagues in the Bahamas to investigate the sedimentary processes at the western escarpment of the Great Bahama Bank, a nearly 800-foot (243-m) vertical undersea cliff. They were particularly interested in how much sediment was being deposited at the base of the bank as compared to the amount being swept into the Atlantic Ocean by currents. As part of their research, they deployed sediment traps along the escarpment, using the small, two-person submersible *Delta*. When they returned a year later to collect the traps and assess the contents, they ran into a slight problem—the traps were gone. They found only one of the sediment traps, about 300 feet (91 m) downslope from its original position.

The reason for their equipment's disappearance became obvious during a subsequent and rather dramatic submersible dive. Bob was the observer and Rich Slater the sub pilot. Rich is well known in the science community as an outstanding sub driver and an endlessly enthusiastic companion on deep-sea dives. He is also the holder of the Guinness world record for the deepest free ascent without support, and more importantly he survived the submersible accident at 225 feet (69 m) that required it. During their sub dive in the Bahamas, when it was time to return to the surface, Rich, as usual, put air into the ballast tanks to start a slow ascent. What happened next took them both by surprise, recounts Bob. "We stopped at 600 feet and then, all of a sudden, the sub started to descend back down, all by itself." They looked out the viewport and could see particles streaming by in the light. A strong vertical current was pushing the sub back down! After descending for about 100 feet (30 m), Rich added more air to the ballast tanks to overcome the current's pull. Thankfully, the sub once again began to ascend. At about 200 feet (61 m) from the surface, they suddenly exited the downward current, and Rich reacted quickly, trying to release enough air from the ballast tanks to compensate for the combined effect of being out of the current and the expansion of the air in the tanks as they rose. But it wasn't quick enough, and the sub rocketed upward, "breaking the surface like a breaching whale." Luckily, no one was hurt, and there was no damage to the submersible.

The presence and strength of the downward current astounded everyone on the research expedition, and it altered their understand-

ing of the processes influencing sedimentation on and around the
Great Bahama Bank. Bob was once again amazed by the power of
the ocean, even beneath the surface. Years later, other scientists
documented flows of high-density water coming off the Bahama
Banks due to cooling by cold fronts or evaporation at high temperatures. This "density cascading" essentially creates waterfalls in
the ocean. Bob and Rich may be the only people to ever have gone
over the falls—undersea!

TIMING IS EVERYTHING

While doing research in graduate school, Bob Cowen, now at the
Rosenstiel School, had a major revelation in the field. It greatly
influenced his career path as a scientist and was due mainly to being
in the right place at the right time, serendipity. In 1983 Bob was
studying the role of sheepshead fish as predators within the kelp
beds of the small islands off the California coast. During his research
he discovered that some sheepshead populations had mostly young
and small fish, whereas other populations were composed principally
of old and larger fish. Based on the differences in size, their role as
predators was clear: the bigger fish could consume larger prey. Less
clear, however, was what had caused the variations in the sheepshead
populations.

Bob had nearly completed his experiments and was planning
one last trip into the field. He had accepted that with only one field
session left it was unlikely that he could unravel the mystery of the
differences in sheepshead populations. As chance would have it,
he was then delayed by about ten days. When he returned to the
field and visited the kelp beds off San Nicolas Island, he got a big
surprise, something he had never seen at the site. There were tiny
sheepshead "all over the place." He collected a few of the young fish
and by extracting their otoliths, or ear bones, was able to estimate
that they were just a few days old. He had stumbled on to what

scientists refer to as a recruitment event. Many organisms in the sea begin life as planktonic larvae that drift in the ocean's currents. Eventually they settle in specific habitats and begin to take on their adult form. At this point they are known as recruits. For Bob the question was why had the recruitment event happened just there and just then? And why, during the previous five years, hadn't he witnessed any other recruitment events at San Nicolas or any of the other islands?

In 1983 a strong El Niño was under way, and in California it periodically caused a reversal of ocean currents. Bob discovered that it was this reversal of currents that had pushed a pulse of water laden with sheepshead larvae into the kelp beds around San Nicolas Island. He also realized that distinct recruitment events such as this could cause the size differences in the sheepshead populations at the various offshore islands. In Bob's case, timing was everything. If he had gone into the field ten days earlier as originally planned, he would have completely missed the recruitment event. Finding the plethora of tiny sheepshead that day not only solidified his research project, but also directed the science he would undertake for years to come. Bob has since focused much of his research on the influence of larval transport and the connections between fish populations.

The strong El Niño of 1982 and 1983 impacted more than just the fish in California. Dick Seymour is a well-known physical oceanographer and coastal engineer at Scripps Institution of Oceanography in California. During the winter of 1982/83, one strong storm after another struck the coast of Southern California. With each storm, the waves pounding the shore got bigger and their periods, the time between crests, grew longer. By March 1983 the waves striking the shore had become the biggest ever recorded in the region—a few even exceeded 46 feet (14 m) in height—and over the winter resulted in damages estimated at more than $100 million.

Dick, along with everyone else in the wave science community on the West Coast, was riveted by what was going on, but at the time didn't know the cause. He suspected that the massive eruption of El Chichon volcano in Mexico six months earlier was causing the wave extremes along California's coast. The eruption had injected enough material into the stratosphere worldwide to increase its temperature by 4°C. Dick knew, however, that his volcano hypothesis would be difficult to prove. There were no other equivalent eruptions at a time when there was also adequate information about the waves in the Pacific. Plus the development of El Niño that same winter was sure to confound the problem. He decided that the first thing he needed to show was that it wasn't El Niño that had caused the big waves.

Using records of rainfall in Southeast Asia and fishing success in Peru, both of which are heavily influenced by El Niño, Dick was able to examine the timing of El Niño events stretching back hundreds of years. Although actual wave measurements weren't available along the California coast till the 1970s, Dick had access to estimated wave data associated with major storms back to the beginning of the twentieth century. He compared the timing of big wave events with strong El Niños and was astonished by what he found. Almost 90 percent of California's really stormy winters during the twentieth century occurred during strong El Niño events. Instead of proving that El Niño wasn't to blame for the big waves, he did just the opposite, sending his original hypothesis out to sea in the process.

SURFERS' DELIGHT

I can't help but tell another of Dick's excellent stories here. In this case it was not the results of his research that were unexpected, but rather the popularity of the data. Dick began measuring waves along the Southern California coast in 1976. At the time he set a wise and prophetic policy for his team: "I made it a rule that the sun could not set on unanalyzed data and that results had to be disseminated rapidly to users such as coastal engineers, other scientists, and educators." Large quantities of data that require analysis can easily pile up to haunt a scientist. And in decades past, the dissemination of research results to a broader audience was not always a high priority.

Dick's group began producing a monthly wave-data report that was mailed out the first week of the following month. When the Internet came online, they quit producing printed reports, instead launching the Coastal Data Information Program on the Web in January 1997. To their surprise, within just a few months their modest server was swamped by thousands of hits every day.

To keep the data flowing, Dick's team had to occasionally replace batteries inside his wave-measuring buoys or deploy substitutes for those that went adrift, causing a brief break in service. Angry e-mails often followed such lapses. Soon they realized where all the traffic to the site was coming from—surfers wanting to know what the waves were doing and if they should call in sick that day.

As their research progressed, wave measurements were made in more locations, and they used a computer model to extrapolate between sites, eventually providing wave data along the entire California coast. They also began to produce three-day wave-height forecasts. The number of hits on their Web site rose to some 85,000 per day. Of course, like all fieldwork, they ran into a few unexpected problems as well, including a strange bias in the data from one particular buoy. It was explained when they discovered a pesky sea lion that had grown fond of hanging out on the buoy. The wave-data program is now international in scope with observations ranging from Mexico to Canada, as well as in Hawaii, Guam, and Brazil and along the east coast of Florida. More than 200 commercial entities download the data every half hour as it is updated. What began as a strictly scientific endeavor unexpectedly became a popular and useful tool for the public, especially the surfing public.

WAVES INSIDE THE SEA

Just as waves propagate along the sea surface, they can also travel through the interior of the ocean—these are called internal waves and usually occur at the interface between two water layers of differing densities. Oceanographer Mel Briscoe provides an interesting and unexpected story about internal waves from a summertime research cruise on Stellwagen Bank, off the coast of Massachusetts.

Stellwagen Bank is a submerged, sandy, kidney-shaped plateau that sits about three miles (4.8 km) north of Cape Cod within a larger, mainly deeper area known as Massachusetts Bay. Average depth on the bank is about 100 to 120 feet (30 to 36 m). In the summer, during hot, calm periods, daytime heating can cause the waters off New England to stratify, with a layer of warm, low-density water at the surface sitting atop a colder, denser layer. Researchers had previously discovered that when the strong tides in the region turn, about every twelve and a half hours, and interact with Stellwagen Bank, they create internal waves at the interface between these warm and cold ocean layers. Once generated, the internal waves propagate to the southwest toward the small coastal fishing village of Scituate, moving at about two miles per hour. Mel was on a cruise to further investigate how these internal waves form and to determine their impact on the plankton and other marine life in the area.

Mel's first surprise on the cruise came when researchers released a neutrally buoyant float and it was promptly sucked downward by the leading edge of an internal wave. The float then popped back up on the other side of the wave. He says, "If that had been a diver, the ascent rate would have been three times faster than recommended to avoid the bends. Submarines have problems with internal waves too, but we had no idea of the magnitude until we saw it with the float." Mel distinctly remembers being able to see the internal waves with the equipment aboard the ship: "We had some new acoustic instruments that let us peer down into the water and actually visualize the waves as they formed, moved, and rolled up like surf and broke." Even more surprising was that the whales in the region seemed to be tracking them as well.

When the tide turned and the internal waves moved from Stellwagen Bank toward shore, whales would be waiting. They were cognizant of a fact the scientists were soon to discover: plankton were being concentrated along the fronts of the internal waves. Each time the tide turned, the whales knew just where to go for a twice-daily buffet of fast food, plankton style.

It was only a seven-day cruise, but Mel says the days were packed and resulted in scientific publications, some innovative ideas, and new friendships. The best part for him was the sense of discovery they had, 24/7; they were doing and seeing new things and connecting physics and biology. He didn't want to sleep and mostly didn't.

Based on the abundance of marine life in the area, its vulnerability, and the region's long history of human use, Stellwagen Bank is now the centerpiece of another national marine sanctuary. The area is an important nursery and feeding ground for many whale species, including endangered humpbacks, northern right, sei, and fin whales. Seabirds are also plentiful, while important fish and invertebrate populations include bluefin tuna, herring, cod, flounder, lobster, and scallops, along with sea turtles. The area is also busy with ship traffic, as mariners make their way to and from Boston. It is hoped that the sanctuary will better enable human use of the region, while helping to protect and sustain its natural abundance.

DISAPPEARING SHARKS

In 2001 Michelle Heupel was studying sharks with marine biologist Bob Hueter at Mote Marine Laboratory. They had spent several years investigating the blacktip sharks in Terra Ceia Bay in the southern end of Tampa Bay, Florida. The blacktips use the area as a nursery grounds. Bob and Michelle tagged hundreds of young sharks, and Michelle used a gridded acoustic tracking system to study the sharks' movements. They discovered that the baby blacktips are born in late spring to early summer and stay for months mainly within the protected, shallow waters of Terra Ceia Bay to avoid the larger, predatory sharks that lurk in Tampa Bay's deeper waters and in the nearby Gulf of Mexico. Around late September to early October, the blacktips begin streaming out from the safety of the bay and migrate south through the Gulf to the Florida Keys.

In mid-September 2001 a strange occurrence took place—all of the blacktip sharks in Terra Ceia Bay suddenly disappeared. Their complete absence in the bay was so surprising that Michelle checked the acoustical tracking system for a catastrophic failure. The equipment was working fine, and within two weeks all of the sharks had returned. Michelle then began searching for an explanation for their unusual behavior.

She discovered that at about the same time that the sharks had left, Tropical Storm Gabrielle was strengthening out in the Gulf of Mexico and heading for Tampa Bay. The storm was nearly a hurricane when it passed quickly to the south of the bay, and it was then that the sharks had started to return. Clearly, the blacktips of Terra Ceia Bay had somehow sensed the approach of the storm and evacuated the shallows, seeking the risky refuge of deeper water. But how did they know to go? Michelle compared wind and rain data and found no correlation to the timing of the sharks' movements. She then examined the record of barometric pressure associated with the storm—the fit was perfect. The sharks had sensed a drop in pressure as the storm approached and instinctively knew to leave. When all was clear and back to normal, they sensed it was safe to return. Coincidentally, at about the same time, other scientists published research suggesting that a shark's inner ear is in fact pressure sensitive. Sharks have long been known to host an array of sensitive, sophisticated senses; from Michelle and Bob's work, add one more and a new way to forecast storm warnings!

ROCKS AND ROLLING

I too have had moments of enlightenment in the field. One occurred during the research for my master's degree, and while it did not lead to an earth-shattering discovery, for a graduate student with preconceived ideas about what I would find, it was an exciting and unexpected result. For students and scientists alike, such moments of insight are a thrilling part of the scientific process, even when the ramifications go no further than their own research.

I was investigating the growth, distribution, and composition of rounded nodules of calcium carbonate found east of Florida's reef tract, at a depth of about 100 feet (30 m). Similar nodules found fossilized in the geological record are typically believed to be indicative of marine environments in which waves or currents are strong enough to cause rolling. I therefore hypothesized that the nodules found off Florida's coast were round because they were intermittently rolled by bottom currents or storm waves.

To map the nodules' distribution, collect samples, and look for clues to the origin of their rounding, I did as many dives as time, funding, and personnel would allow. I found that the nodules were made up of accreting layers of coralline algae and encrusting fora- minifera, both of which secrete a thin, crustlike skeleton of calcium carbonate. In some areas there was evidence, such as scours and sediment tails in the shadow of larger rubble, indicating that currents periodically swept the seafloor and were rolling the nodules. But in other areas, I found no such evidence and a surprisingly lush growth of macroalgae. What was turning the nodules in these areas?

While slowly ascending to the surface one day at the end of a dive, I happened to look down at the seafloor. The light and visibility were just right, so that from above I could see a strange pattern below—there appeared to be a network of long sandy trails amid the macroalgae. I distinctly remember stopping midwater and pondering the unusual markings on the seafloor, but because we had limited time at that depth, I could not go back down to investigate further.

I examined the mysterious trails on the seafloor during my next dive, discovering to my surprise that within each, just beneath the surface, was a seabiscuit, the echinoderm *Meoma ventricosa*. These short-spined creatures are plowlike deposit feeders, meaning that they crawl through and ingest the sediment to feed on the organic matter within. They were also acting as undersea bulldozers, pushing aside the nodules in their path and exposing the undersides to the growth of encrusting organisms. While I originally believed that physical processes were solely responsible for the turning that created a rounded shape, I came to realize that biological influences played a role as well. Again, not a discovery that rocked the scientific world, but a good example of how a moment in the field can change our perspective and render preconceived ideas erroneous, at least partly so. It also illustrates how biology, geology, and physical influences interact within the sea.

From groundbreaking finds to moments of personal enlightenment, the unexpected results we encounter in the field are of the utmost importance in ocean science. They provide the insights we cannot envision, program, or predict in the office. They are the discoveries that often come by happenstance, by simply being in the right place at the right time.

Underwater photographer Porter Watson calls filming tiger sharks up close and personal an adrenaline-pumping treat. Photo courtesy of Porter Watson, TropicSea Visions.

The abundance and diversity of marine life on a healthy coral reef is a true ocean wonder. Photos Ellen Prager.

An army of sea urchins mows down the seagrass in Florida Bay during a 1997 bloom. Photos Ellen Prager.

Bob Halley and Kim Yates of the U.S. Geological Survey create their own portable undersea incubation chamber to run twenty-four-hour experiments on water chemistry on a reef. Photos courtesy of Kim Yates, U.S. Geological Survey.

(*Opposite*) Sea lions present unexpected obstacles: *top*, a tug-of-war with Josh Feingold's camera in the Galápagos, and *bottom*, hanging out on a wave buoy. Photo on the top Ellen Prager. Photo on the bottom courtesy of Dick Seymour.

A typical seascape in St. Croix before the mass mortality of the black spiny sea urchin, *Diadema antillarum*, in 1983. Photo courtesy of Bob Carpenter.

Undersea drilling of coral cores in St. John, U.S. Virgin Islands. Photo Ellen Prager.

Navigating a precarious dive entry for a student field trip in St. Croix. Photo Ellen Prager.

The volcanic landscape and island of Bartolomé in the Galápagos. Photo Ellen Prager.

Helping park rangers repair a path up the side of a volcano on Bartolomé. Photo Ellen Prager.

Sea lions playing nearby in the Galápagos—a pleasant distraction. Photo courtesy of Josh Feingold.

Sea Education Association ship SSV *Westward*. Photo Ellen Prager.

A wave crashes over the bow of the SSV *Westward*. Photo Ellen Prager.

Sealab's assistant Tuffy helps out undersea. Photo courtesy of Dale Anderson.

4 *Forces of Nature*

ON CALM DAYS, THE OCEAN CAN BE SMOOTH as glass, as inviting as a cool, crystal-clear lake on a hot summer's day. On other days, the sea can rapidly morph into an angry tempest of towering waves and whipping winds that is anything but inviting. Under these conditions, one questions the sanity of doing science at sea. In the field, safety must always be the top priority, but even with the best of plans and preparation, we can still learn a thing or two about nature's rapidly changing moods and the power of the ocean and atmosphere.

RACING A HURRICANE

In 1992, on my very first teaching cruise as a chief scientist for SEA, I encountered a raging gale, sharks, and a hurricane—*none of which appeared in my contract!* We departed from Woods Hole, Massachusetts, aboard the SSV *Westward* in early October for a six-week voyage through the open waters of the North Atlantic south to the Caribbean. The *Westward* is a 125-foot schooner originally designed for circumnavigating the globe as a private yacht. It was later refit as a sailing-school vessel and outfitted for oceanographic research. At the helm of the *Westward* was experienced mariner and nautical science faculty member Phil Sacks. Soon after leaving port, we encountered gale-force winds and about 20-foot (6-m) seas. The

Westward performed well, sailing smoothly through the strong winds and large waves. Most of the students, however, were beleaguered by seasickness. Luckily, I am rarely susceptible to seasickness (knock on wood), yet it was rough enough to make me wonder what I had gotten myself into this time—*little did I know.*

Soggy and worn out after the gale, we sailed south into the Sargasso Sea and calmer waters. The sun was finally shining, drying everything out and warming our spirits. We held classes and conducted oceanographic sampling, and the students stood watch, learning how to sail and navigate. After being chilled and damp for days on end, the students pleaded for a swim call in the warm waters of the Sargasso. Ever mindful, Phil was reluctant to allow a group of twenty-four students to jump, splash, and make nothing short of a commotion in an area of the ocean where food is scarce—shark food, that is. The Sargasso Sea and many open ocean areas are nutrient-poor and therefore considered relatively barren. Having dealt with sharks in coral reef settings (plenty of shark food there), I didn't initially share his misgivings—evidence of my inexperience in the open ocean. Phil begrudgingly agreed to a swim call, ordered the ship hove-to (sails set so it stays relatively stationary), and put two of the crew up in the rigging on shark watch. I decided to join the students in a quick dip.

The water was incredibly calm, clear, and blue. The serenity of the sea was soon disrupted, however, as nearly all of the students jumped in and splashed about. There was also a deep, vibrating thump that occurred each time a student leapt from high up in the rigging. I quickly got a decidedly uncomfortable, wary feeling—the hairs stand up on the back of your neck kind of feeling—and climbed out of the water. Minutes later there was a holler from above as several sharks were sighted on the surface swimming toward the ship. We reacted immediately, getting everyone out of the water as calmly and as quickly as possible. The sharks lazily approached and then simply swam silently off in another direction. It is unclear what would have happened had the students still been in the water, but I for one didn't want to find out. I learned that day to trust Phil's judgment even more than I already did and to worry about sharks during swim calls in the open ocean.

At this point in the cruise, while most of us aboard were thankful for the improving weather, Phil was wary, watching the clouds with a more experienced eye. As he and I later sat below in the aft cabin listening to the daily marine forecast, the National Oceanic and Atmospheric Administration (NOAA) reported that a tropical storm had formed southeast of Bermuda, about 300 miles (480 km) away. The storm was predicted to move north and then slowly recurve to the east. Phil calculated that if the storm stayed on its forecasted track and we kept to our current heading to the southeast, our closest point of approach would be about 120 miles (190 km). He noted, however, that overall it was not a good scenario—it had all the ingredients for the making of a late-season hurricane.

I went up on deck to teach the oceanography class, while Phil stayed below to confer with the mates about the impending weather. Phil later briefed everyone to explain the possibilities and the seriousness of the situation without causing undue alarm. He explained that given the predicted track and the development of the storm it would be best to stay on our current heading and try to avoid the strongest winds.

Everyone onboard worked diligently to prepare the ship by securing sails, equipment, tools, the galley, and personal items—everything that could be lashed down was, and tight. Safety lines were also rigged throughout the ship to allow the crew to move about without having to unclip their safety harnesses. The winds were now blowing a moderate 20 to 30 miles per hour and the seas were about 10 feet (3 m).

That evening we felt as prepared as possible and continued heading southeast while anxiously awaiting the next weather update on the marine radio. The good news—the storm was continuing to move north. The bad news—it had intensified. The storm's winds were now steady at over 63 mph with gusts up 74 mph, and it was predicted to become a hurricane within twelve hours.

Tension onboard was high, as we knew the storm was strengthening and its path could easily shift. We continued to make way motor-sailing with a small amount of sail up to help reduce the roll of the ship and to push us as quickly as possible to the southeast.

At midnight Phil and I sat anxiously by the radio listening to the next update; it was Advisory 3 for what was now Hurricane Frances. The storm was packing steady winds of 74 mph with gusts up to 92 mph, and, worse, the predicted track had shifted to the east. The forecast also called for further intensification. Phil and I immediately went to the chart table and plotted the new track of the hurricane against the ship's cruise track. We were now on a collision course with the hurricane!

Because hurricane forecasting is an inexact science, we decided to call the National Hurricane Center via the marine radio. First time through, the marine operator got a recorded message saying to call back during regular business hours—*what?* We convinced the operator that there must be another number and to try again. Once connected, the forecasters told us that they were confident in the storm's predicted track. Explaining where we were and who we were (a ship of ten crew and twenty-four undergraduate students out in the open ocean directly in the storm's path), we asked again if they were sure—really sure, this was no esoteric question. Again they assured us of their confidence in the forecast.

Given the change in the storm's track and intensity, we needed to make a decision—continue trying to pass the storm to the east, but risk being caught in the dangerous right-hand quadrant of the hurricane (in the Northern Hemisphere the combined direction of the wind and movement of the storm make this the most powerful and dangerous part of a hurricane) or try to cross the storm's path into a more favorable, navigable quadrant, but risk being closer to the eye. It was a very difficult decision. Phil says it was "one of the toughest he had ever had to make as a captain."

Phil decided that it would be best to cross the hurricane's path, get on the more favorable side, and put as much distance as possible between the storm and us. So we headed to the west, while Hurricane Frances barreled to the northeast—the race was on. As the wind strengthened and the seas grew, we used both wind and diesel power to run from the storm. Phil recalls that in the early morning hours the barometer fell fast and "spectacular bolts of lightning lit up the otherwise pitch black sky." Only mates and one student at a time were allowed up on deck to steer and keep an eye on the rigging and sail.

Later that morning, being the curious scientist that I am, I couldn't resist the opportunity to see what it was like in the midst of the storm. I put on my foul-weather gear and a safety harness, and went up on deck to observe the conditions. Rain driven horizontal by the wind struck my face like pellets, and there was a strange hue to the air, with lightning and moisture providing a pinkish tinge. It was difficult to estimate the height of the waves, as the howling wind was shearing off their tops, leaving streaks of white water across the ocean's surface. According to the ship's log, we were experiencing 60 to 70 mph winds and 20-foot (6-m) seas. Waves broke over the rails, washing across the deck, yet the *Westward* continued to ride well.

Eventually and without serious incident, we made it across the hurricane's path and into the more favorable side of the storm. We passed about 120 miles (190 km) from its eye with just minor damage and one injury aboard. One of the mates was literally tossed from her bunk when the ship took a sharp roll on a wave. For most of us, it was just bumps and bruises. Our sister ship, the SSV *Corwith Cramer*, which was about twenty-four hours behind us, was less fortunate. They passed closer to the eye of the storm, incurring more damage to the ship and at least one injury that initially looked quite serious.

When we reached our first port of call on the cruise, the small, wonderful island of Bequia just south of St. Vincent, everyone was glad to be in a protected harbor and spend a little time on dry land. Being prepared and having the appropriate knowledge, skills, and equipment are essential while in the field. On the ocean, things can happen fast and decisions have to be made quickly. The experience with Hurricane Frances forever solidified my respect for the sea, well-crafted ships, the potential power of a storm, and the wise judgment of experienced crew. When I hear reports that a hurricane poses no threat as it remains at sea, I often wonder if there are ships in harm's way.

Some of the details about our experience on the SSV *Westward* in Hurricane Frances come from an excellent account written by Phil and published in 1997 as part of a chapter on hurricanes and hurricane maneuvering in the book *Auxiliary Sail Vessel Operations* by G. Andy Chase.

SCIENCE VERSUS SAFETY

Facing storms at sea can be especially precarious when working with expensive scientific equipment and when time is limited. As bad weather looms, it can be very difficult to decide when and if to call a halt to operations. Al Hine, a well-respected geological oceanographer at the University of South Florida, with a vast amount of experience working in the field and on large oceangoing ships, humbly tells of one situation that provides a frightening and illustrative example.

Al was the chief scientist overseeing a geophysical survey in the Gulf of Mexico aboard Duke University's R/V *Cape Hatteras*. The ship was some 120 miles (190 km) offshore when the captain informed Al that a strong cold front was approaching. As always seems to happen, it was just at the critical point of their seafloor survey. Al, as the chief scientist, had to decide if they should continue collecting data or hunker down for the impending storm. As the air temperature began to plummet, the wind rapidly strengthened and the seas grew, but Al decided to continue surveying, just a little longer.

The *Cape Hatteras,* like many other large oceangoing research vessels, is outfitted with a large open deck or fantail at the stern for deploying or towing heavy equipment. Working on the back deck during operations, even in calm conditions, can be dangerous. But in a storm, it is the place of nightmares.

With conditions worsening quickly, Al and his team realized that they needed to secure their supplies on the ship's fantail and pull in the costly geophysical gear they were towing. At this point, however, the growing waves were causing water to wash across the deck, and with the rolling of the ship the rails were being submerged. In this sort of sea, just walking without a firm handhold becomes a challenge. Add rain and growing darkness, and the stern deck was not a place anyone wanted to be. The need to secure their gear became more urgent when Al looked aft through a porthole and saw that some of their equipment had broken loose. Given Al's experience, the captain reluctantly gave him permission to go out on deck to secure the gear, though he advised against it. The captain then worked to orient the ship into the waves to provide the most stable position possible, while Al chose the most experienced graduate student in the group to help. They decided on a plan, made their way aft, and ventured out into the storm. Al describes the experience: "We both fell as we approached our unsecured gear and were soaked by water rushing across the deck. We stood up and began retying our lines, having to grip each other just for stability. At one time, the water racing across the fantail was more than knee deep." At that point he realized the true danger of their situation and wished they had rigged up safety lines to ensure that neither of them were washed overboard.

Al and his student were able to secure the gear and make their way safely back into the ship's main science laboratory. In hindsight, he sees that going out on deck during the storm was unwise: after all, the equipment was replaceable. He realizes now that he continued surveying longer than he should have—not an uncommon decision among scientists—and that he should have heeded the captain's caution more seriously and at least asked for help from the crew. Al recalls that during the experience he and his student were no longer adviser/student, but two people facing a more dangerous situation than they had anticipated. The graduate student went on to become a professor at a major university and Al continues to go to sea, but he, like most of us who have had similar experiences, has a healthier respect for the ocean, especially in storm conditions.

RUN AWAY!

It is easy to conceptualize the threat posed by hurricanes or major fronts in the open ocean, but even small, intense squalls relatively close to shore can be dangerous. During the summer in south Florida, afternoon squalls with strong winds, torrential rains, and frequent lightning are the norm. In Florida Bay the small mangrove islands that dot the region are about as tall as the surroundings get. This brought little comfort to Bob Halley and me those squally summer afternoons in the bay in a small boat with a long metal rod (aka lightning attractor) that we used to probe the sediment depth. When storms developed, we typically watched them warily and headed to shore as they approached, while they were still a safe distance away.

One day, however, our strategy failed. The squall moved in much faster than we expected, and before we could run away, its strong winds and black skies were upon us. Driving rain poured down, and lightning flashed as the sharp crack of thunder rang out. We ran the boat toward shore as fast as possible, bouncing roughly atop the steep chop and white-capped waves that had enveloped

Florida Bay. As we tried to protect ourselves from the stinging rain and wind, I remember looking over the side, and I swear the seafloor was visible in the troughs of the waves. Along with wind strength and the distance over which the wind blows, water depth is a key factor in how big waves can get. During that squall the waves may have been literally as high as they can get in the bay's shallow water. After a bone-rattling boat ride, we made it to shore and took refuge. Like Al, we too should have called it a day sooner, but were enticed by the desire to get as much data as possible during our limited time in the field.

HIGH-ENERGY EVENTS

In the language of ocean science, storms and the associated sea conditions they produce are referred to as high-energy events. Or in my field-relevant terms, kick-the-crap-out-of-the-scientist weather. The truth is, scientists have a poor understanding of exactly what happens in high-energy events in the ocean because it is so difficult to get good data during them. This is particularly true in environments such as the surf zone or on the top of coral reefs where it is simply too rough for firsthand observations or most instruments, which cannot take the physical beating inflicted by smashing waves, strong currents, and blasting winds. I can personally attest to this fact.

In the early 1980s, I assisted marine ecologist Bob Carpenter in surveys of the black spiny sea urchin *Diadema antillarum* on a coral reef in St. Croix. We typically anchored our small boat in the calm, shallow waters behind the reef, donned scuba gear, and swam through a narrow cut in the coral to the fore reef (the steeper front of the reef facing the open ocean). On calm days this was perfectly safe and an easy way to get across the reef without damaging the coral or ourselves. Plus it saved us a lot of time. We avoided having to motor a much greater distance to a larger cut in the reef, go through in the boat, and then double back on the outside of the reef to the survey site.

Our shortcut through the reef worked great until one day when a small but powerful squall developed offshore while we were diving. We were so engrossed in counting *Diadema* that we didn't notice the strengthening of the wind or thickening cloud cover. I don't remember if it was thunder or the sound of large waves pounding on the reef crest that got our attention, but it definitely became clear that a strong squall was nearing and we needed to get back to the boat and to the safety of shore. Being caught on the surface in a squall would mean having a steel tank on our backs in lightning or getting smashed onto the reef by the strengthening wind and waves. Neither was such a good option. Bob and I decided our best bet was to try to find the cut in the reef and swim across.

As waves break on a reef, water initially surges forward and then there is a strong flow backward. This powerful two-way surge combined with very shallow water and dangerously sharp coral at a reef's crest can be tricky in low-wave conditions, but in high-energy events it should, without question, be avoided.

We swam as far up onto the reef as possible while still in control, grabbed on to stop our motion, and searched for the narrow swim passage through the coral. In the white water from breaking waves and as our bodies flipped forward and then back in the surge (don't try this at home), we couldn't see the reef cut. Without much of a choice, we waited for what seemed like a lull in the wind and waves, let go of our handholds, and hoped that our wet suits would take the brunt of the beating. The strong surge pushed us forward and then sucked us back. The trick was to ride, as calmly as possible, the water's forward motion across the reef, trying to steer your way through the coral, and then kick like hell when the backward surge struck. It probably took only a few minutes to get across, but of course it seemed a lot longer.

Our experience is illustrative of how coral reefs incur damage during a storm. Waves crashing on a reef in high-energy events break off large pieces of coral. Once loose, rocklike corals roll and tumble about, smashing and breaking other corals. Evidence of this bowling-ball effect can often be seen on a reef after a storm.

In the field, vigilance and constant monitoring of the surrounding conditions is critical, even for veteran field-goers, as things can change surprisingly fast. It is also important to inform those who are on the surface standing watch in a boat to notify divers if bad weather approaches. In stormy conditions and rough seas, getting into a pitching boat with lots of gear on can be difficult and downright dangerous. Being banged on the head by a dive platform swinging violently up and down is called—a concussion.

An interesting note about *Diadema* and coral reefs. Prior to 1983 on coral reefs throughout the Caribbean, the Bahamas, and Florida, these black spiny sea urchins were incredibly abundant, as many as three per square foot (thirty per square meter)—that's wall-to-wall sea urchins and a spiny hazard to divers. Their long, thin black spines are tipped with toxin and can be quite painful when stuck in the skin—an all-too-frequent occurrence, given their plentiful populations at the time. By the way, there is some science behind the idea that urinating on a sea urchin wound will help. The spines are made of calcium carbonate, and the acidity of urine acts to dissolve the tips of any spines left in the skin (I'd go with vinegar). During 1983 and 1984, some 95 to 98 percent of the *Diadema* throughout the tropics and subtropics of the Atlantic suddenly died. It was shocking to go out on a dive one day where the *Diadema* were abundant and seemingly healthy and the very next day find the seafloor littered with black spines and tests.

Scientists now believe that a disease carried by ocean currents caused the massive mortality of *Diadema*. Researchers also think that the loss of *Diadema* from reefs has contributed in many areas to the rampant growth of algae, which can overgrow and smother corals. Black spiny sea urchins are one of the reef's best vacuum cleaners, crawling over its surface and consuming massive quantities of algae. In some areas *Diadema* populations have started to slowly rebound on their own, but with algal overgrowth becoming a serious problem in many areas (overfishing of herbivores and nutrient pollution also contribute to the problem), researchers have also been experimenting with rearing them in the laboratory and transplanting them onto coral reefs.

Bob Carpenter continues his research on coral reefs out of California State University, Northridge, though these days he spends more time in the Pacific than in the Caribbean. Looking back, he says that on most days the conditions in the Caribbean were benign compared to the Pacific. Diving on the outer reef off the island of Moorea, he typically encounters 6-to-10-foot (2-to-3-m) swells breaking on the reef—on a normal day. These large swells generate currents across the back reef, which are so strong that Bob says working there is akin to horizontal rock climbing.

A NATURAL VORTEX LAB

From aboard SEA's SSV *Westward* comes another forces-of-nature story, this time from marine scientist Duane De Freese, now at Hubbs-SeaWorld Research Institute in Orlando, Florida. In 1976 he was aboard the *Westward* as an apprentice. They were two days out of Key West, Florida, sailing in heavy weather when ahead loomed a fast-building frontal system complete with towering stratocumulus clouds—an indicator of strong vertical motion in the atmosphere. With all hands on station preparing for a blow, they steered a course south of the squall line. As they skirted the storm, a waterspout dropped from the black sky above. Then another waterspout dropped down, and within an hour and a half they saw a total of eight waterspouts within just a half-mile of the ship (approximately). Duane is convinced that every student and each member of the crew aboard *Westward* that day understood, as never before, the intimate connection between the atmosphere and ocean. Over three decades later, the incident was still vivid in his mind when he read the following quote in *USA Today* from Joseph Golden, a senior NOAA scientist: "The Florida Keys are the greatest, natural vortex lab in the world."

Over a decade after Duane's experience, I was on an SEA ship when a waterspout nearly ripped across the deck. The crew and students responded quickly, making an unplanned tack and avoiding it by tens of feet. As the waterspout passed to our starboard side, we got a close-up view of the strong rotating winds that whip water up and off the ocean's surface into what is essentially a tornado at sea.

Onboard SEA's tall sailing ships, the crew is especially wary of fast-developing squalls, now more than ever in light of the tragedy that befell the *Pride of Baltimore* in 1986. The 137-foot schooner was sailing some 240 miles (385 km) north of Puerto Rico when a sudden burst of strong wind, also known as a downburst or white squall, struck without warning. Eighty-mile-per-hour winds reportedly hit the ship, put it over on its side, and sank the *Pride of Baltimore* so fast that there wasn't even enough time to radio for help. Three crew and the captain were lost, while the surviving eight men were cast adrift in a life raft for four days until a passing freighter picked them up. Though such incidents are rare, remembering that they can and do happen is important when working at sea.

FOG AND FIRE

For Peg Brandon, now a professor of maritime transportation at the Maine Maritime Academy, her first voyage as a captain with SEA in 1986 also made a lasting impression. She was aboard the 134-foot brigantine SSV *Corwith Cramer* heading north from Woods Hole, Massachusetts, toward the coast of Nova Scotia on the first leg of a six-week cruise. Then came the fog. For the first two weeks of the trip, day after day, they were completely enveloped. To navigate, Peg had to rely solely on the electronic equipment aboard. Fog can be a true hazard, especially in areas with heavy ship traffic, marker buoys, and small islands. And while 1986 is not that long ago, at the time they did not have access to the high accuracy afforded by the Global Positioning System (GPS)—LORAN-C and radar were the name of the game. There were no collisions or even close calls, but the experience made Peg realize how reliant she normally was, particularly in the coastal zone, on navigation by sight. I sailed with Peg in the early 1990s on that same northern track, and we too hit a blanket of fog over Georges Bank. While I was nervous, Peg took it all in stride, navigating with ease purely by the electronics onboard.

As a captain, Peg points out that GPS has been a true revolution in the maritime industry, making navigation, search and rescue, and daily operations at sea safer and more efficient. Yet she fears that while advances in technology are of great value in science and maritime transportation, they can also divorce us from the benefits of simple observation. On SEA ships the students are taught to navigate by the stars and watch the clouds and sky for indicators of impending weather. Students collect buckets of seawater to measure ocean temperatures, in the process looking out over the sea, noticing its color, the wind, and wave state. They also learn that even with the lowest of low-tech methods, things can go wrong—such as forgetting to tie off the line to the bucket before throwing it in the ocean. Now students have access to GPS, weather faxes, satellite imagery, and flow-through seawater systems that automatically measure ocean temperature and a slew of other parameters. With the ever-growing reliance on technology are students, scientists, and mariners still making the simple observations that help us understand the ocean and atmosphere? In the fog at sea, we have to rely on instruments rather than the power of observation, but on clear days has our vision become clouded by technology?

In the field, Peg has also witnessed a somewhat unusual and mysterious force of nature. She was aboard the SSV *Corwith Cramer,* this time sailing near the Bahamas. Peg remembers it being a calm night with lightning far off in the distance. The ship's huge mainsail suddenly began emitting a loud crackling sound, and a greenish glow appeared in the rigging above. It was St. Elmo's fire, a phenomenon that occurs when the atmosphere becomes electrically charged and voltage builds up in a ship's sail and rigging before being released as a "glow discharge." Mariners historically considered St. Elmo's fire a good omen or protection from a storm. For Peg the phenomenon was entrancing, yet at the same time made her wonder, "What the heck am I doing on deck?"

WILD WATERS

Part of being a graduate student is working for and assisting your advising professors. A benefit of this is the opportunity to tag along on research projects other than your own or to go on professional field trips as cheap labor. Sometimes, however, we get more than we bargained for.

In the mid-1980s my adviser led a geological field trip to the Bahamas aboard a trawler that had been converted into an all-purpose research vessel for hire. My fellow graduate students and I went as the help. In addition to standing watch, cooking meals, and assisting in the field, that meant, to our surprise, sleeping on deck or the floor under a table in the main cabin (the paying customers got all of the actual beds aboard).

The trip began uneventfully. We departed from the southeast coast of Florida, crossed the Gulf Stream, and arrived in the Bahamas. There we visited numerous sites to view a diversity of carbonate sediment environments and to discuss their origin and how they related to the geology of ancient settings. We had two unexpected encounters while in the Bahamas, one with a Bahamian patrol boat looking for drug smugglers and the other with possible modern-day pirates. The pirates were fortunately scared off when a savvy member of our crew climbed to the bow of the boat, quite obviously flaunting a rather large gun. With scientists working the world over today, researchers and their sponsors have to be more wary of security issues. In 2001 a large research vessel working off the coast of Somalia was attacked by pirates; luckily those aboard were able to fend off their attackers and no one was injured. The incident, however, heightened the ocean science community's awareness of security issues while in the field.

The real story from our trip began on the return voyage to Florida. We left the Bahamas and crossed the Gulf Stream on a beautiful, calm day with swells traveling east to west, the same direction we were motoring. As we cruised toward Florida, the swells grew and the boat began to roll steeply from side to side. The captain assured us that the trawler was designed for this and if necessary, to go up on one side, but always come back—a fact that many of us aboard didn't find all that comforting. The renovated trawler also had flopper-stoppers, essentially long booms that can be lowered to the water line for added stability. Just a short time later a wave hit us, and the boat rolled so far over that several of us slid precariously across the deck and a few pieces of unsecured gear were lost over the side. The captain again assured us of the boat's stability.

When we arrived at the Florida coast, the captain headed toward Haulover Inlet on his way into the Intercoastal Waterway. Haulover Inlet, as I later learned, is known for its treacherous currents. I was at the stern of the boat, about to climb up a ladder to the upper deck. Another graduate student, Lenore Tedesco, was behind me, and several of our guest luminaries were sitting on a bench against the back rail of the boat.

As the boat entered the inlet, it hit the currents and waves just right so that the trawler began to lean to one side. Leaning turned into rolling, and I grabbed the ladder as the deck literally dropped from under my feet. Lenore grabbed my waist, and the professors at the stern grabbed the rail as water rose over the now nearly vertical deck. The boat rolled all the way over onto its side with both propellers completely out of the water. It stayed that way for what seemed a very, very long time, though it was probably only thirty

seconds or so. It was long enough for the captain's daughter to climb
to the bow preparing to jump and for me to contemplate how to
get off the boat without being trapped underneath if it capsized.
The boat then slowly, painfully slowly, rolled back.

No one was physically hurt in the incident, but there was plenty
of mental distress to go around. The captain was shaking, ready to
sell the vessel. My adviser had been trapped below when the boat
rolled. He came up white as a ghost, gripping a bottle of scotch.
Our near capsize was as close as I ever want to come to a tragedy
at sea. The captain subsequently put about 2,000 pounds of lead
in the trawler's keel and made many more, less eventful, voyages. I
suspect a factor contributing to the incident was that the boat was
top-heavy, overloaded with gear on the upper deck. Experience
can make us less attentive to important details, though sometimes
nature just throws us a curveball and there is little to do other than
react accordingly.

BIG WAVES AND STRONG CURRENTS

Two ocean phenomena that neither my colleagues (that I know of)
nor I have encountered in the field are tsunamis and rogue waves.
The 2004 Indian Ocean tsunami off Sumatra tragically illustrated
the destructive power of this force of nature. Scientists are working with government agencies and local communities to better
prepare and plan for tsunamis and to establish an improved warning
system in areas most at risk. Fieldwork is an essential component of
tsunami research, especially after one has occurred. Scientist Costas
Synolakis at the University of Southern California's Tsunami Research Center is part of an international team that deploys shortly
after a tsunami has occurred. They go to the impacted area to
investigate the details of how and where the tsunami struck and the
cause or triggering event. This information is vital to create better
warning systems, to rebuild more safely, to improve education,
thereby preventing future casualties, and to predict what areas are

most at risk in the future. As for rogue waves, researchers have recently shown with satellite imagery and radar that they occur more frequently than previously believed and most commonly develop in areas where strong currents interact with opposing wind and waves.

One side note on dangerous currents in the sea—each year about one hundred people die when caught in rip currents, swift jets of water flowing offshore from a beach. Unfortunately, people caught in rip currents often try to swim back to the beach directly against the flow, tiring quickly. Because these flows are usually relatively narrow, if caught in a rip current, you should swim along or parallel to shore and then back to the beach.

DANGEROUS MARINE LIFE

During a reconnaissance trip in Fiji, while walking in the shallow water of an exposed reef flat, I asked my guide about sea snakes—the deadly poisonous kind. He replied, as I remember it, "Oh, one just swam by, but don't worry, they have very tiny mouths and are extremely docile. But you should be careful where you step due to the many deadly stonefish and cone shells in the area." With planning, experience, and some common sense, scientists can avoid most encounters with dangerous marine life, though bad luck does take its toll every so often. For the most part, scientists go to sea to observe the ocean or organisms in their natural state, so great care is taken not to disturb or harass marine life. It is also important to remember that even the most docile of creatures will react aggressively if provoked, and that wild animals are just that, wild animals.

The stories shared in this chapter are not meant to imply that the ocean is to be feared or a dangerous workplace to be avoided. They are meant rather to showcase the respect the ocean warrants, the experience that is needed to safely work at sea, and what we can learn when Mother Nature flexes her oh-so-powerful muscles. My colleagues and I have also recounted some of our own mistakes and the somewhat dangerous situations we have found ourselves in, in this chapter and the next, so that others can learn from our experiences and go to sea as well prepared as possible.

5   *Overcoming Obstacles*

IN PREPARING FOR FIELDWORK AT SEA, scientists must be optimistic, yet learn to adhere to a useful motto—what can go wrong probably will. For the ocean is an unforgiving mentor. Ever-changing sea and weather conditions are the norm. Mechanical and electrical failures are likely, and equipment may be lost. If there were a field guide to ocean science, it would say to also consider uncooperative marine life and the constant threat of bad luck or human error. Furthermore, by its very nature, the ocean makes research difficult, hiding what we want to see and to study. Its vast, wet, and deep expanses limit our access and require creative means to tease out the secrets that lay within.

The obstacles that arise in the field may be foreseen or unexpected. Experience in the field teaches us that simplicity is to be valued, and that duct tape and tie wraps are as essential as water, sunblock, or a GPS. Some problems can be solved easily, while others test our will and ingenuity. We learn to adapt both in terms of our own needs and those of our science. Yet by overcoming obstacles in the field, we become stronger, more committed, and versatile, and it can lead to advances in technology or knowledge. In the end, the challenges fought in the field will either enrich our success or leave us wanting a new career choice.

I asked my colleagues what frustrates them the most about working in the field. You might expect replies about weather delays, mechanical failures, incompetent assistants, or even sharks. Those, however, are considered a part of the adventure and the challenge of working in the sea—they go with the territory. Researchers find limited and inadequate funding, the burdens of bureaucracy and risk management, and the increasingly exorbitant amount of time that must be spent writing proposals, filling out paperwork, and participating in administrative meetings much more frustrating and difficult to deal with. And as George Matsumoto of the Monterey Bay Aquarium Research Institute sees it, it's not only a challenge to find funding, but surprising that more money is not directed toward basic ocean research and exploration.

One big problem is simply that the ocean is difficult and costly to study. Our access to the sea is limited both in space and time. Imagine how much simpler it would be if we could just look into the ocean from the surface to see what lies at the seafloor or what creatures were swimming within. Marine biologist Sonny Gruber of the Rosenstiel School has been studying sharks for more than thirty years. From his early work on shark anatomy and physiology to his current ten-year effort to better understand the life history of lemon sharks, Sonny has become one of the world's foremost experts on sharks. Yet over decades of study, he has witnessed sharks feeding undisturbed in their natural habitat just a few times.

Limited access to the sea and its creatures has spurred an exciting, relatively new field of research that relies on tags tracked by satellite or undersea acoustical arrays. Results are exposing, for the first time, the wide-scale travels of some of the ocean's larger organisms, such as tuna, whales, sea turtles, billfish, and even great white sharks. Barbara Block of Stanford University, a pioneer in the use of satellite tags, and her colleagues tracked one great white shark over 2,200 miles (3,800 km) from the Farallon Islands to Hawaii. This type of work is particularly important for understanding where such creatures reside, feed, and reproduce: information that is essential to conservation efforts. But even with satellite tracking, it remains extremely difficult to get long-term records that allow us to delineate monthly, seasonal, and yearly patterns or how organisms interact with each other or the environment, especially when it comes to some of the sea's smaller creatures.

A NO-NET PLANKTON TOW

While studying the ecology and early life history of fishes, Bob Cowen became frustrated with traditional sampling methods. He was investigating the early life history of billfish, such as swordfish, sailfish, and marlin, where adult populations spawn, and how the larvae of these fish are dispersed. Bob and his students relied on the traditional technique of towing a large net behind a ship to locate and collect fish larvae. Net tows can damage planktonic organisms beyond recognition, however, and the results reveal little about how they are distributed or oriented in the water. Here was the crux of the problem: Bob was particularly interested in the spatial distribution of the organisms in the water, and their orientation to prey and each other.

Unwilling to accept the status quo, Bob worked with his students and colleagues to solve the problem by creating the next-generation plankton sampler—a high-tech in situ imaging device that can be towed behind a ship. The new imaging system not only leaves fish larvae and other planktonic organisms alive and intact, but it also records with great resolution their fine spatial distribution, orientation, and abundance over large distances. Bob likens the camera to a copy machine, which scans images line by line. Make that a camera that records 36,000 times per second underwater while being towed at up to five knots. The result is a 6-inch- (15-cm)-square image over the length of a tow. Bob calls the early results from the imaging system nothing short of amazing. Each frame captures a picture of all the larvae present in their natural position in the water, and because most are transparent, it even reveals their internal structure. In the process of overcoming an obstacle, Bob and his colleagues have created a new window through which to view and study the sea's planktonic populations.

QUANTIFYING THE GLOW

When Edie Widder began studying bioluminescence in the ocean, she wondered why so little was known about it. One reason soon became clear: at the time there was no way to measure or quantify bioluminescence. Edie and her colleagues set about to create the technology needed. They developed a large screenlike device that when pushed in front of a submersible maps the distribution of light emitted from animals that it comes into contact with—the device became aptly known as the SPLAT CAM.

Edie and her team recently deployed a new system in the Gulf of Mexico to record deep-sea animals attracted to bioluminescence. It uses an autonomous, programmable, battery-powered camera with red-light illumination that is invisible to most deep-sea creatures. The camera is paired with an innovative electronic lure that imitates the light display of a deep-dwelling jellyfish. Edie believes that these jellyfish use bioluminescence as a burglar alarm. If a predator attacks, the jellyfish emits light to lure in larger creatures that will see its attacker as a better, heartier meal. Just eighty-six seconds after the electronic jellyfish lure was turned on for the very first time, at a depth of about 2,000 feet (610 m), a 6-foot- (2-m)-long squid came into view. Scientists are still studying the images of the squid, as it is a species that nobody had seen before. The deployment of the new camera and electronic lure system in the deep sea was not without its challenges. It was discovered early on that the camera was tilted upward, offsetting the image from the desired viewing area. Edie's team came up with a quick and ingenious solution. They found an aluminum ladder on the ship and after getting permission from the captain took it apart to create a device that put the camera at a better angle for viewing—it is now a standard part of their deployment and is known as the CLAM, or cannibalized ladder alignment mechanism. A new version of the camera and lure system is under development for a long-term deployment in the deep waters of Monterey Canyon and will include video streaming to shore.

A GALÁPAGOS ADVENTURE

Not all hindrances to research or fieldwork lead to advancements in knowledge or technology; sometimes they are just challenges that must be overcome in order to obtain data, any data. Then again, sometimes we just want to make it back to shore safely.

In the late 1980s I went on a research expedition to the Galápagos Islands that, upon looking back, now seems like a bad B-movie—with marauding sea lions, sharks, lots of large hairy spiders, a dangerously incompetent crew, and even thieves. I was there for two months assisting renowned coral biologist Peter Glynn, from the Rosenstiel School, and his graduate student at the time, Josh Feingold. They were studying the long-term impact of the 1982/83 El Niño on the region's coral reefs. To this day, it was one of my most memorable, in a good way, adventures in ocean science.

The Galápagos Islands lie some 600 miles (965 km) west of Ecuador, straddling the equator. They are hot, dry volcanic islands bathed in surprisingly cool waters that flow north from the Antarctic or upwell from the ocean's deeper reaches. The islands' location and ocean conditions as well as their isolation from the continents have produced a unique and utterly fascinating combination of species and many endemic creatures, found nowhere else on Earth. From a scientist's perspective, it was oddly intriguing to be studying warm-water corals yet have cold-water animals, such as sea lions, fur seals, and penguins, swimming nearby.

Our main goal was to survey coral reefs that had been studied prior to the 1982/83 El Niño to assess its impact and the corals' prospects for recovery. In our dive planning, however, we neglected to consider curious or aggressive sea lions. During our dives off the island of Santa Cruz, we frequently encountered female sea lions and pups that were extremely playful and overtly curious. They seemed to consider our underwater survey gear as potential toys or fodder for a friendly game of tug-of-war. I have to admit, compared to the other problems I've experienced in the field, this one wasn't so bad. Though when your dive buddy is in front of you and something unseen tugs at your fin or pulls at your hair, it can be rather unnerving. Sometimes we were simply distracted from our scientific duties, watching as the sea lions nearby enjoyed a game of catch; they would pick up a small rock from the bottom, toss it into the water, and try to catch it before it sank to the seafloor.

The male sea lions or bulls, on the other hand, were dangerous; one had recently bitten a park ranger, causing serious injury. They are large, strong, and have very big, sharp teeth (not too mention stinky breath from eating all that fish). Bull sea lions are also territorial and can be quite aggressive. At one survey site, a none-too-happy large bull chased us out of the water. On other occasions, I have been chased out of the water by sharks that were a little too curious for comfort and once by dolphins when we happened upon a pod during mating activities. There is something to be said for common sense when it comes to fieldwork and safety.

We faced more serious challenges during a research cruise from the Charles Darwin Research Station on Santa Cruz to a few of the other islands in the archipelago. At the time, there was just a small tourist industry in the Galápagos, and few boats were available for charter that a scientist could afford. Peter therefore rented what seemed a sparse but adequate local boat for our needs.

The week began with a diesel spill, which we later found out had rendered all of the water onboard undrinkable and the boat itself slick and stinky. Once under way, the smell of diesel and rough seas left several of our research party miserably seasick. For Josh and me, sleeping was difficult as our bunks were in the bow of

the boat directly under the anchor and chain locker. Each bounce of
the boat resulted in a bone-jarring thump directly over our heads.
After diving all day in cool water, we had only sickly sweet pink
soda or beer to drink. Potable water, as it turned out, wasn't the only
thing the ship's crew had neglected to pack. Days into the cruise,
away from any means of resupply, food began to run short, with
the exception of rice, spaghetti, and a few moldy potatoes. The
mate's fishing prowess, or lack thereof, resulted in an interesting if
not unappetizing dinner of moray eel served atop a bed of spaghetti
mixed with rice. The mate's small-boat skills roughly equaled his
fishing ability. He nearly ran over us while we were diving and then
left us stranded after a drift dive, so that we had to make a long swim
back to the boat with all of our gear.

When the cruise ended, we were grateful to be back at the
Darwin station, until we realized that Peter's bag with his passport,
money, camera, and all of his data had mysteriously disappeared.
He was overwrought, not so much about his passport, camera, or
money, but the years' and years' worth of data that were in his
missing notebook. And he had no backup of the data. The next
day we searched frantically for the bag, perplexed as to how it was
lost. Josh and I even dove around the boat, thinking that maybe
we had dropped it while offloading. A few days later, after being
pressured by the owner of the boat, the oh-so-wonderful mate
confessed to stealing the bag. He was promptly arrested, and we
heard that he was to be deported from the islands. He had slyly
slipped the bag under the floorboards of the small boat that he
shuttled us to shore in. Peter got his notebook back, and we all
learned a few important lessons—check out a boat and its crew
before chartering it (if possible), assess the supplies they have
garnered for your trip before leaving the dock, and always, always
make a backup of your data and store it somewhere safe before
going into the field.

Compared to the charter boat, our base of operations and accommodations at the Darwin station were luxurious. Josh and I had a small room with two beds, a bathroom, and access to a small boat or panga, scuba tanks, and a compressor for air fills. Unfortunately, we also had a cold, sulfur-smelling shower and a lot of very large spiders. At night the ceiling of our room was literally covered with big, hairy spiders, but what really got under our skin was that when we woke up in the morning, they were all gone. Where did they go? Josh reminds me that the spiders did have a redeeming quality—they kept the cockroaches in check. We did have one true luxury just a short hike away—fresh homemade ice cream at the one hotel on the island.

One of our treasured items during those long days in the cool water was a sun shower, a large, thick plastic bag that we filled with water each morning and hung up in the sun to warm while we were out diving. One day when we returned to shore shivering, anticipating that short, deliciously warm shower, we discovered, to our great dismay, that our sun shower had been stolen.

One of our survey sites was just off the small island of Bartolomé. By coincidence the Galápagos National Park Service was going there to repair a path for tourists up the side of an old volcano. They invited us to hitch a ride. We joined several rangers on a park-service boat for the overnight voyage to the island. The captain was a proud, kind man who spoke little English. Assuming that we were a couple, he graciously offered us the largest bed on the ship for the night—the captain's bunk. It was a bit awkward since we were not a couple, but we accepted anyway, not wanting to insult the master of the ship. When we arrived at Bartolomé the next day, the captain shuttled us to shore, curious about our work. We landed on a narrow beach, but our survey site was on the other side of the island, over a steep sandy hill. Josh and I put on our wet suits, hoisted our scuba tanks over our backs, and hiked up the hill carrying our equipment. It was a short but hot hike, to say the least. Upon reaching the top, with the captain at our side, we were confronted with an interesting view. Below us was the small embayment we needed to survey, but so was a small school of circling Galápagos and hammerhead sharks. The captain found this quite amusing, pointing to the sharks and us, the crazy gringos. To his surprise, we promptly hiked down the other side of the hill and got in the water. Josh claimed ladies first, though I suggested that since it was his research project he should take the lead. The water was a bit murky, and I looked over my shoulder more than a few times. The sharks stayed away, uninterested in the crazy gringos, but our adventure on Bartolomé was just beginning.

After we completed our work, a boat from the Darwin station was to pick us up to ferry us back to Santa Cruz. We finished a bit early and decided to help the park rangers while we waited. Their repair efforts entailed carrying heavy bags of cement, large buckets of water, and hefty logs up the side of a sandy volcano. Since we didn't want to seem like feeble Americans, both of us set to work carrying loads as similar to what they were carrying as possible. It was hot and steep, and the loose sand made it so that with every step up we slid partway back down. By the time the small boat arrived to shuttle us back to Santa Cruz, we were especially grateful and totally exhausted.

We had a long boat ride back to the lab, and the driver informed us that given the distance and choppy seas, we would need to ride in the tiny bow compartment where the anchor and line were stored. We thought he was kidding. He wasn't. Josh and I crawled into the compartment and curled up next to the anchor line with our heads resting (soon to be banging) against the hull. It would seem a truly uncomfortable position for the long ride back, especially as the boat bounced over each and every wave, yet we both fell sound asleep—a testament to our exhaustion. I did have a bit of a headache, however, when we arrived in Santa Cruz.

Sometimes in research the obstacles are just as much mental as they are physical. In the Galápagos there were times when research was not going well and motivation was hard to come by. Comforts were lacking, and it was often difficult to get things done. Repairs required scavenging for supplies or tools, and back then we had no means of communicating with anyone outside of the station. It was tough, remote, and, overall, wonderful—of course, such experiences are often remembered with rose-colored glasses.

I have since returned to the Galápagos several times. There is now an airport instead of a hut on the short runway, as well as paved roads, several hotels, and many boats for tourists, even a small cruise ship. The wildlife and volcanic scenery remain some of the world's most spectacular; however, problems have arisen because of the growing population. In particular, overfishing has decimated the lobster and sea cucumber populations and threatens other fisheries in the region. Unfortunately, with little other means to support themselves, local fishermen have turned to exporting their catch, in some cases illegally. Hopefully, profits from responsible tourism will be shared with the local community and, along with better governmental policies, can be used to protect this unique and fascinating place.

Peter and Josh (now Dr. Feingold) found that the warming wrought by the 1982/83 El Niño devastated the corals in the Galápagos, killing 95 percent of those present. By the early 1990s, however, corals in the region were beginning to recover, but at a slow rate. These corals are acclimated to relatively cool water, and only a few degrees' increase in ocean temperature, as happens in the eastern Pacific during strong El Niños, is enough to kill them—sadly this may foreshadow what global warming will do to coral reefs around the world. The 1982/83 El Niño has also turned out to be an important watershed event for scientists studying the ocean. It revealed the true impact and important influence of El Niños on the marine world, and many scientists began to study the phenomenon as an essential part of their research. We now know that, in physical oceanography, in marine biology and geology, and in coastal climatology and weather, El Niños are an important driver of change in the sea and on the planet.

THE ENEMY WITHIN

From a captain's perspective, Phil Sacks notes that sometimes scientists can create their own obstacles as well as challenges for others. He recalls a teaching/research cruise during which they encountered several fishermen adrift off the coast of Nicaragua. While he supervised assistance, worried about the very real potential for piracy, and saw to everyone's safety, amazingly, the chief scientist aboard kept pressuring him to continue bottom sampling on a nearby bank.

When working in the extremes of the Antarctic, science can be especially difficult and dangerous. Phil recalls scientists aboard a cruise in the Antarctic that wanted to collect a three-ton, algae-ridden ice block so that they could extract the algae and feed it to krill in the ship's wet-lab aquaria for onboard experiments. The collection of a huge ice block from the middle of a broken ice flow in the dark, freezing Antarctic winter, however, was easier planned than done. They were eventually successful in getting the huge chunk of ice onboard the ship, but the experiment proved futile, as the krill died anyway. On another cruise in the Antarctic, large seas prevented landing at a shore-based station and almost cancelled an entire field season. In planning for another attempt to land, Phil told the scientists involved to unpack *all* noncritical gear. When they finally made it ashore, Phil was livid because lives had been put at risk and, "I discovered about 25 percent of the remaining *critical* supplies were booze. An overturned boat, in Antarctic waters, even wearing exposure suits, could easily have meant death." Safety in the field must always be of the highest priority, even at the expense of science or our own creature comforts. Yet, when engrossed in collecting data or when experience and familiarity breed complacency, it is sometimes easy to neglect safety and even common sense.

MISTAKES HAPPEN

By the time I entered graduate school in Miami in 1985, I had more field experience than most students, but in truth I still had a lot to learn. Two incidents during the fieldwork for my master's degree are illustrative of what not to do.

One day after making all of the extensive arrangements necessary for a day in the field, I, along with my partner for the day, drove from Miami to the Florida Keys, towing a small boat. I was feeling fortunate that day to be accompanied by an expert boat mechanic, diver, and all-around field guru. All was not so rosy, however, when we got to the boat ramp in the Keys and I discovered that my field guru had forgotten an important piece of equipment—*the key to the boat*. But he assured me that all was not lost as he could hot-wire it. This seemed a reasonable, if not resourceful, solution. He got the boat running, and we made our way several miles offshore, nearing my research site. We then discovered a problem: hot-wiring a boat drains the battery even while running. Alas, now a good distance offshore, once the motor was shut off we were unable to restart it, and since the radio also ran off the boat's battery, it was dead as well.

We had no choice but to wait for another boat to happen by while suffering the steamy heat of a calm, sunny, summer day in the Florida Keys. Unfortunately for my colleague, he is also one of those masochistic souls who go to sea for a living, yet are prone to seasickness, even in the calmest of conditions—any anger I felt at that point was replaced by empathy for his misery. We eventually caught the attention of a fisherman passing nearby. He had no radio though, and his boat was even smaller than our own. After an unsuccessful attempt to jump-start our battery, we pleaded for a tow to shore. He agreed only after we sweetened the deal with $50 and

an anchor—*bribes may be needed in fieldwork.* The moment a towline
was secured, he began motoring toward shore. We proceeded to
watch in horror as he aimed straight for a reef. His small boat would
pass easily over the shallow coral, but for us it was questionable.
We shouted, jumped up and down, and did a lot of arm waving, to
no avail. We then scurried onto the very bow of the boat, hoping
to act as a counterweight, raising the stern and propeller as high as
possible in the water. The boat zigzagged as it approached the reef
with us cringing, crouched at the bow. Somehow we made it over
without hitting the coral and eventually reached the shore. I swore
from that day on that if it were my field project, I would always take
responsibility, or at least check on, any critical piece of equipment,
no matter how small.

On another day of fieldwork for the same project, I faced the
consequences of a different and potentially more dangerous mistake.
In fact, this was one of the scariest experiences I've had in the field,
and it was not because of sharks or storms or boat malfunctions—it
was the result of my own complacency. As I explained earlier, for
my master's degree I conducted research in the deep waters off
Florida's reef tract. As part of the project, I did as much scuba diving
as was needed and possible. The water depth involved, typically
about 100 feet (30 m) or greater, made research a bit more difficult
as we had very limited bottom time each day. Along with making
observations and collecting samples at the seafloor, I decided to set
up a growth-rate experiment. I enlisted several expert divers from
the University of Miami to help. The idea was to put small, dyed
limestone blocks on the seafloor, which I could collect months later
to determine the growth rate of encrusting organisms.

On the day of our dive to set up the experiment, there was a stiff
northward current running on the surface—probably a meander of
the Gulf Stream. Based on previous experience, we assumed that at
the bottom the current would be minimal. We hung several scuba
tanks off the boat to be used, if needed, during a safety decompres-

sion stop at the end of our dive. Three of us then geared up and got in the water, while one person stayed on the boat for safety.

At the surface we swam a short distance to the anchor line carrying our gear and fighting the current. We then had to literally pull ourselves hand over hand down the anchor line to get to the bottom. At the seafloor it was a bit murky, but as we suspected, there was little current. All of us were very experienced divers and quickly set about our assigned tasks. In the poor visibility I briefly lost sight of my colleagues, but knew that they were working nearby. Mistake number one. Then I got an unpleasant surprise—the airflow through my regulator suddenly became stiff. Knowing all too well what this meant, I immediately checked my pressure gauge. I was essentially out of air. In all my years of diving, I had never had this happen. I almost always came up with more air left than anyone else, and I religiously checked my gauges. Whether it was fighting the current or a leak or a low fill at the start, I don't know, but I found myself 110 feet (33 m) below the surface and out of air. To make matters worse, my buddies were intent on their work and just out of eyesight—I couldn't signal them to buddy breathe or that I was in trouble. At this point I was sucking the remnants of air from the tank and physically fighting the urge to hyperventilate. The only choice was to make an emergency ascent to the surface. I knew that it was crucial to ascend as slowly as possible, all the while exhaling to prevent getting the bends or, even worse, an air embolism, both of which can lead to serious injury or even death (more in the next chapter). It was one of the most difficult things I have ever had to do, both mentally and physically. To make matters worse, when I neared the surface, I could see the spare tanks hanging off the boat, but although I am a very strong swimmer, because of the strong current I couldn't get to them.

When I reached the surface, I took several giant breaths of air and hollered to the person aboard the boat as I was rapidly dragged north by the strong current. Unfortunately, he did not expect us

up for at least ten minutes and didn't see me or hear me yelling. I
tried to relax, breathe slowly, and wait for my colleagues to realize
my predicament. Luckily, because of the experience of my crew, it
was only a few minutes before the other divers came up and the boat
was motoring my way. I was shortly thereafter plucked, gratefully,
from the sea. Fearing the effects of my rapid ascent, once on the
boat I sucked down the oxygen from a bottle we had aboard for
emergencies (oxygen can help flush excess gas from a diver's system).
I was lucky that day and suffered no ill effects from the incident. I
have never since been complacent about my air supply, before or
during a dive. As one of my colleagues says, a series of small errors
can cascade into a serious problem and lead to disaster.

By the way, my growth-rate experiment was lost. At the time
we didn't have access to GPS to get an accurate position at the surface
overlying the experiment. We did mark the site with submerged
buoys (buoys on the surface marking research or equipment are often
stolen or lost), but with only an approximate location at the surface,
the water depths involved, limited dive time, and the flat, essentially
featureless terrain of the seafloor in the area, we were unable to
locate the experiment again. I used the growth rates suggested in
the scientific literature for my research, but even now would love
to find that site—because, much like a fine wine, the data would
be even better with age.

SPUN AND SLIMED

In the mid-1980s marine geologist Denny Hubbard (now at Oberlin College) was studying the influence of land development and
construction on coral reefs in St. John in the U.S. Virgin Islands. As
part of his work, Denny examined the skeletal banding in corals to
determine if sediments released during development had stressed
the nearby reefs. Each year, reef-building corals typically produce
two skeletal bands, a dense or heavy band and a less dense or light
band, much like annual tree rings. An X-ray of a thinly sliced core

through a coral's skeleton is used to view these bands and allows a scientist to determine a coral's age and look for pattern disruptions that may be indicative of environmental stress. Some skeletal bands will also fluoresce under a black light, indicating that a high amount of organic matter has been incorporated into the skeleton. This can occur as a result of increased runoff during storms or possibly from the release of soil or sediment due to construction. Scientists can also examine the isotopic signature or chemistry of the skeletal bands to estimate past ocean temperatures or salinity. The first step, however, is to drill a core from a coral's skeleton.

I joined Denny in St. John for a week of drilling coral cores. At the time the technology to do so was relatively new and primitive, and left us battling a force known as torque. The drill consisted of a large, cylindrical, steel core barrel with a cutting bit on the bottom. It was powered by a motor sitting in a small boat overhead and controlled by a throttle on a horizontal bar atop the core barrel. The problem was that, as the core barrel and drill bit spun, cutting into the coral, torque caused the horizontal control bar—and the divers holding on to it—to also spin. We spent hours underwater kicking against the spin or wedged in against the reef trying to prevent it. Denny informs me that while drilling in Puerto Rico a few years later they got smart, and to combat torque they replaced divers with a piece of rope tied to a convenient bit of "dead" reef some 20 feet (6 m) away. Seems so simple now. In St. John, however, after days of drilling underwater we were exhausted, and our hands seemed permanently curled into the shape of the drill handle. Additionally, the large head corals we were drilling responded by excreting copious amounts of mucus, which they normally do to slough off sediment and other particulate matter that settles on top of them. In other words, we got seriously slimed by the coral. After drilling, we filled the core holes with underwater cement, which enabled the coral to regrow over the surface, thereby preventing any further damage due to coring. Given the challenges involved, at the end of a week in the field, the cores we were able to extract were prized possessions, the truly hard-won fruits of our labor.

Methods for the drilling of corals underwater have improved greatly since my foray into the field with Denny. Coral skeletons have also proved fruitful in terms of research, providing a history of El Niños in the Pacific going back hundreds of years, a record of past hurricanes in the Florida Keys, and a useful means to help decipher climate change. As for Denny's research in St. John, the results were inconclusive. He had excellent land-use records for comparison, but the data from the skeletal banding showed little correlation as well as unexpected variability in growth rates, even at a single site. Denny came away wary about using skeletal banding in corals for anything more than calendars in conjunction with other analyses. Our experience is illustrative of what can happen in the field or in science in general. Arduous efforts may result in just a few precious samples and sometimes results just don't pan out, but that's part of the process, and when such endeavors are successful, it is all the more gratifying.

⁓⁓⁓⁓⁓⁓⁓⁓ The challenges faced in the field can seem endless in variety and perpetual in nature. Yet with experience we become better at anticipating and solving problems at sea. We learn what can be overcome and how much we can personally accomplish, even under difficult circumstances. And on the days when success does come, there is great satisfaction, personal reward, and, sometimes, really good science.

Given the problems and difficulties that are often faced in the field, one might wonder if a career in ocean science is worth it. I believe I speak for most of my colleagues in saying that the answer is a resounding YES! We wouldn't trade our chosen professions or the adventures we've had for anything. Our experiences and what we have learned from them far outweigh any of the discomforts or frustrations that we have had to endure. And when those moments of wonder, discovery, or scientific enlightenment do happen, it makes all the obstacles we've dealt with seem inconsequential and simply fodder for funny stories over cocktails. Going into the field isn't easy—little in life is. But surely it is worth it, both personally and for the invaluable knowledge we gain about the sea.

6 *Living and Working Undersea*

IN THE FIELD THERE IS ONE THING ALWAYS in short supply—time. Scientists crave it. They need time to make observations, to collect samples, and to conduct experiments, but getting extended time on the sea or beneath the waves is costly and problematic. After all, we are air-breathing, land-dwelling creatures that must rely on technology to enter the watery world of the ocean. The more time we spend in the field, however, the more we can learn. For projects that require extended time underwater or access to the deep sea, it is especially difficult.

Maritime archaeologist Nicolle Hirschfeld from Trinity University in San Antonio, Texas, understands the problem of time all too well. It took her and an international team of experts eleven long years to survey and excavate a 3,000-year-old shipwreck lying off the west coast of Turkey. The wreck was relatively deep, 150 feet (45 m), in a remote location, and divers had limited time each day during a short diving season. They were further challenged by huge undersea drifts of sand and work that required meticulous care. Once uncovered, artifacts had to be described, sketched, and photographed, and their location precisely measured before they could be removed. Many objects had become cemented to the seafloor, and as Nicolle puts it, "chiseling them out demanded patience with steady hands and thousands of delicate hammer strikes." Divers also spent many hours delineating where goods had scattered away from the wreck and lay strewn across the seafloor.

The team's determination and years of effort on the shipwreck paid off, opening a new window into the seaborne trade of the region in 1300 BC, roughly the era of the famous Egyptian pharaoh Tutankhamen. Found among the ship's cargo were ten tons of copper ingots, most of which were mined on the island of Cyprus. There were tin and raw materials, such as glass ingots, hippopotamus and elephant tusks, exotic woods, and ostrich eggshells, for the making of luxury items. Almost 400 shipping jars were discovered that once carried wine, olives, or resin for the manufacture of perfume or incense. There were also precious goods, including jewelry made of gold, silver, and amber, along with more pedestrian items, such as tableware and glass paste beads. Nicolle is passionate about the wreck and its remaining mysteries: "It is not always easy to separate cargo from personal effects, and the basic questions of the ship's route and who was on board and the purpose of the journey remain unanswered."

Underwater, time is precious, especially for scientists. After years of doing fieldwork on scuba, worrying about time and air, for me one experience stands out above all others—living underwater. Living in an undersea research laboratory or habitat allows a freedom that most divers can only dream of. When living at about 45 feet (14 m), researchers can dive 6 to 9 hours at depths between 45 and 95 feet (14 and 30 m). And when combined with access through a habitat's windows, living undersea affords an unparalleled view of the ocean and its creatures 24/7. Being inside a habitat, akin to a fixed-in-place underwater mobile home, is a lot like being in the reverse of a fish bowl—a human bowl, where the fish roam free and the humans are entrapped within. But the humans periodically get to escape on dives, so it is not true captivity, though anyone susceptible to claustrophobia may argue that point. For me, a stay within an undersea habitat means I am no longer just a momentary visitor to the marine world; I am living within and among its inhabitants.

Twice now I have lived in the *Aquarius* undersea habitat, about four miles off Key Largo, Florida. Many of my experiences while living underwater have become indelible memories. Take our daily commute, for example, as we swam to and from the *Aquarius* to work sites on the nearby coral reef. Powered by our legs and fins, we passed coral formations that became like familiar buildings along a frequently traveled highway. I grew accustomed to seeing a pair of butterfly fish that loitered about a large barrel sponge and watched for a big moray eel that inhabited a specific hiding hole. When conditions changed, I saw the reef's creatures respond. In the late afternoon as daylight waned, fish hovered noticeably closer to the reef, soon to seek nighttime refuge within its interstices. When high seas created surge at the bottom, fish again took to the reef's crevices for protection from the ocean's flow. Under the glare of the habitat's lights, I even saw a barracuda snatch a midnight snack—a tasty red squid. And when a barracuda trailing a long, tangled fishing line, with a hook embedded in its mouth, hovered near a viewport—we sent help. A member of the surface crew dove down, and while the barracuda stayed surprisingly still, he removed most of the line. Throughout the mission, that same barracuda frequently came by as if to say thanks. As a diver and someone endlessly curious about the sea, I relish the unique perspective that living underwater allows. As scientists, my colleagues and I accomplished work that would have taken us many weeks had we been diving from the surface. This is the beauty and the advantage for science that living underwater affords—time.

THE PROBLEM OF PRESSURE

The advent of scuba diving was revolutionary for scientists studying the ocean. For the first time they could swim among the sea's creatures for extended periods and view the ocean firsthand. But even so, scuba diving has its limits. The problem is pressure. At the surface there is one atmosphere of pressure from the weight of the overlying air. With depth in the ocean, the pressure increases due to the weight of the overlying water. With every 33 feet (10 m) of water, there is another atmosphere of pressure. In others words, at about 33 feet down in the ocean, there is two times the pressure at the surface, at 66 feet (20 m) down three times the pressure, and so on. Pressure in most of the ocean, which averages about 12,000 feet (3,650 m) in depth and reaches more than 35,000 feet (10,900 m) in the Pacific, is enormous. For scientists who study the ocean's deepwater realms, the extremes of pressure dictate the use of specially designed, multimillion-dollar submersibles or ROVs. But even in relatively shallow water, the pressure is potentially dangerous, even deadly.

As we breathe air from a scuba tank and dive deeper into the sea, the increased pressure causes our blood to absorb more gas. We must get rid of this excess gas before returning to the lower pressure at the surface, or face serious consequences. Consider a can of soda or a bottle of fine champagne. When either is sealed, relatively high pressure keeps excess gas (carbonation) dissolved in the liquid inside. In a clear container, few bubbles are readily apparent. Upon removal of the cork or opening of the lid, the pressure inside instantly decreases and the excess gas in the liquid expands rapidly, forming bubbles that rise to the top. In the case of champagne, this can occur explosively, shooting the cork across the room. A diver's body at depth can be likened to a champagne bottle that we don't want to uncork.

Years ago, the navy developed rules for safe scuba diving based on the ability of a person to return to the surface with minimal excess gas or bubbles in their system. For instance, a diver can go to 60 feet (18 m) for about sixty minutes and then safely ascend directly to the surface, going slowly and exhaling on the way up. If a diver stays longer or goes deeper for an extended period of time, he or she cannot return directly to the surface without bubbles forming in the blood, which can cause serious injury when lodged in joints, or worse if they enter the lungs, spine, or brain. This is known as the bends or decompression sickness and in severe cases can result in paralysis or death. To avoid the bends, divers must follow the rules of depth and time, or go through decompression (spending additional time at specific depths while ascending) to eliminate the excess gas in their blood.

The potential danger of getting bubbles in our blood limits the time a person can spend underwater without decompression, thus restricting what most scientists can accomplish in any given day. There is a way to extend your bottom time, however—just don't come up. If divers stay at depth for about twenty-four hours or more, their body becomes saturated with gas; this is known as saturation diving and allows divers to extend their time on the bottom by worrying about decompression later. After saturation, divers must go through seventeen hours of decompression to safely return to the surface.

Once, while I was living in *Aquarius*, a doctor made an underwater house call for a routine checkup and to do a little demonstration. He took a sample of my blood while in the habitat and then brought it directly to the surface. The blood bubbled so violently that it shot the stopper out of the vial—a powerful and sobering illustration of what would happen if any of us, once saturated, made a beeline to the surface without going through decompression. Before scientists begin living underwater in *Aquarius*, they must spend a week in training, learning how to deal with emergencies and to react to problems underwater without following their natural inclination—to go to the surface.

SCIENCE AND THE INFAMOUS PARROTFISH

Before I actually lived underwater, my first brush with undersea habitats was in St. Croix in the 1980s when I was a support or safety diver for *Hydrolab*. *Hydrolab* was a small cylindrical habitat, 16 feet long by 8 feet in diameter (about 5 m long by 2 m in diameter). It was originally deployed off West Palm Beach, Florida, before being relocated to the U.S. Virgin Islands. Four scientists lived inside *Hydrolab* each mission, but there were only three bunks—we often thought that Hydro-can was a more appropriate name. As a support diver back then, my job was to track the scientists when they were diving outside of the habitat, to supply them with full scuba tanks and additional gear as needed, and, oh yes, to be an undersea waiter bringing meals down in a large pressurized, water-tight metal pot.

When you're living underwater, food inexplicably tastes strangely bland. At *Hydrolab* this was exacerbated by a cook who was surprisingly skilled at sucking the flavor out of food. We regularly potted down such tasteless treats as "dead bird in a bag" (chicken), "pressed pink, white, and brown" (deli meats), "pureed green" (vegetables), and "roast brown" (an unknown species of cooked meat). But once, when a dignitary was in the habitat, I distinctly remember diving down a pot containing a specially prepared lobster dinner. After we passed the pot through *Hydrolab*'s hatchway, we often stood in our wet suits with full scuba gear on, waiting to bring it back, empty, to the surface. Sometimes the aquanauts would show their appreciation with a tip, just as other diners do—in our case it was a tasty chocolate cookie. Looking back, it was a sort of reward I guess, a bit like those trained marine mammals that do tricks for a treat—*hmmm*.

To track the scientists when they were diving outside of *Hydrolab*, we became especially adept at following their bubbles on the surface. We could communicate to them without having to dive down by using a hydrophone or underwater microphone that we kept on our small boat. On one night mission, a long excursion dive by the scientists stretched into the early morning hours, which left us at the surface tired, cold, and ready for bed. We put the hydrophone in the water and whispered the virtues of hot chocolate and a warm bunk. Shortly thereafter, the diver's lights began moving toward the habitat. On my very first mission as a support diver, I was diving in the dark, late at night, when the staff quietly slipped the hydrophone into the water. As I maneuvered in the surrounding silent and inky blackness to grab a set of empty scuba tanks at the seafloor, I suddenly heard music—the ominous *"dunt, dunt . . . dunt, dunt"*—from the movie *Jaws*. Practical jokes and lighthearted humor often took the strain from what could be physically demanding and, at times, stressful situations. From my time as a support diver at *Hydrolab*, I learned an enormous amount about diving, safety, and fieldwork. And one mission in particular had an important influence on my decision to pursue a career in science.

I worked at *Hydrolab* the summer following my junior year in college. At the time I was contemplating science as a career, but was unsure if I had what it takes to be a "scientist." Part of my uncertainty was probably due to the pervasive stereotype of scientists as intense, unsociable people who are singularly dedicated and have an extraordinary intellect, yet without much of a sense of humor or personality. I was pretty sure that wasn't me (and as I now know, not many of my colleagues).

While I performed my duties as an undersea gopher at *Hydrolab*, I also paid close attention to what each team of scientists was investigating and how they conducted their research—I was an eager science slave whenever possible. On one mission the scientists were studying the role of parrotfish in the recycling of nutrients in coral reefs. Parrotfish are abundant and important herbivores on many coral reefs, except where they have been overfished. Princess parrotfish are particularly beautiful, medium-sized, turquoise fish decorated with touches of yellow, green, and purple. Their most distinctive feature, however, is their rather large, protruding buckteeth, which are used to scrape algae off the reef's coral and rubble surfaces. When parrotfish are dining nearby, it is often possible to hear a symphony of underwater munching and to see the white scars they leave behind on coral heads. While scraping off algae, these voracious grazers also ingest pieces of the corals' calcium carbonate or limestone skeleton. The old adage of what goes in must come out then comes into play—so parrotfish are also prodigious defecators, the true poop machines of the coral reef. The nutrients and sand in their wastes are released back into and thus recycled within the reef ecosystem. Much of the lovely, fine white sand in coral reefs and on tropical beaches that we love so much? Parrotfish poop.

Given the abundance of parrotfish and their propensity for eating and defecating, scientists have long been interested in better understanding the role they play in nutrient recycling in coral reefs, a highly productive ecosystem that typically thrives in nutrient-poor waters. To investigate this question, the team of scientists at *Hydrolab* was attempting to quantify the contribution of parrotfish to nutrient recycling—a fancy way of saying they were collecting parrotfish poop for later analysis at a shore-based laboratory.

Imagine several highly trained and respected scientists on scuba chasing parrotfish around the reef with small plastic bags attempting to collect clouds of excrement. Yes, this was real science, and for me it was a wake-up call—if collecting parrotfish poop was science, I could do it.

BUILT ON THE PAST

Modern habitat programs, such as *Hydrolab* and its larger, more advanced successor *Aquarius*, have had exceptional safety records and supported hundreds of successful research missions. Like much of science, their achievements and progress have been built on the successes and failures of the past. Steven Miller, formerly the director of the *Aquarius* program, provided me with some interesting highlights from the history of undersea habitats.

People who live underwater are called aquanauts, the ocean equivalent of astronauts. The first true aquanauts, however, were goats. In the 1950s and '60s, navy captain George Bond, widely considered the father of saturation diving, led a project to investigate the effects of increased pressure on humans. For reasons of safety, however, some of the early testing was done on goats (not to worry, they were treated well and safely accomplished their mission). Navy divers later successfully spent twelve days in a chamber at a pressure equivalent to that at a depth of 200 feet (60 m). These early experiments opened the door to underwater living and the construction of undersea habitats.

The legendary ocean explorer Jacques Cousteau built one of the first and most picturesque undersea habitats, nicknamed Starfish House for its shape, having five sections branching out from the center. Divers lived underwater in Starfish House for four weeks at a depth of 36 feet (11 m). It was said to be quite comfortable; the undersea residents reportedly drank wine with dinner, smoked cigarettes, and had a parrot living with them. Drinking of alcohol and smoking are definite no-nos in modern habitats. A documentary film about Cousteau's program even won an academy award in 1965. At the time, Cousteau remarked on undersea living, "The hazards were great and were exceeded only by the challenges."

Indeed, those who strove to build the first undersea habitats encountered a whole host of problems, and even today, living undersea is far from easy. The U.S. Navy's first undersea laboratory, *Sealab I*, sank twice and filled with water before a successful launch in 1964 in the Bahamas. A tropical storm then halted the *Sealab* mission after only eleven days, although it was supposed to have lasted for three weeks. Then there was the "Tilt'n Hilton," *Sealab III*, which operated off La Jolla, California, in 1965. It got its nickname because, after settling on the ocean bottom, the habitat developed a serious lean to one side. Overall, the *Sealab* program proved a success with three consecutive ten-day missions undersea. The aquanauts at *Sealab* even had a special underwater assistant, Tuffy the dolphin. She helped with navigation in murky water and delivered tools and supplies to divers working at sites distant from the lab. Her pay? Fish, of course.

The first undersea habitats to be operated over a longer period of time and support scientists doing research were *Tektite I* and *II*. Scientists used their time in *Tektite* to investigate the surrounding coral reefs and to study fish. The longest mission ever conducted in *Tektite* was sixty days off St. John in the U.S. Virgin Islands in 1968. At the time, undersea living had been the exclusive domain of men, but when a woman applied for the program, the managers decided it was time for an all-woman team (today mixed-gender missions are common). The first all-woman aquanaut team, led by renowned underwater explorer Sylvia Earle, completed a mission in *Tektite* and was later celebrated with a ticker-tape parade in Chicago. Sylvia and her scientist colleagues must still wince, however, at the nicknames given to them by the journalists at the time—aquabelles and aquababes.

Soon after *Tektite*, *Hydrolab* was built. It also had its share of problems in the beginning, including one 25-mile (40-km) trip out into the Gulf Stream after breaking loose in a storm. Shortly after, a heavy concrete base was attached to the lab to secure it to the seafloor. From the early 1970s to 1985, over 180 missions were conducted in *Hydrolab* in the Bahamas and later in St. Croix.

One person who was instrumental in the *Hydrolab* program and many of the advances made in modern undersea living is my friend and colleague, diver and scientist Bob Wicklund. He and his wife, Gerri, ran the *Hydrolab* program in the Bahamas and have been actively involved in undersea habitats ever since. For Bob, the 1960s were the true heyday of undersea habitats as the premise of humans living and working in the sea had created passionate interest all over the world. People at the time dreamed of extended stays in the ocean to study its wonders, to farm fish from submarine-based homes, and to live among gardens of coral. There were primitive undersea systems being built in seventeen countries including the

United States, Australia, Germany, Canada, Great Britain, Japan, and Italy. Overall, some fifty undersea habitats were constructed, but when the practical realities of underwater living were realized, just about all of them were relegated to the scrap heap, *Hydrolab* being one of the exceptions.

Bob's appetite for underwater living was first whet when diving from a small submersible at 120 feet (36 m) and realizing that he wanted more time among the sea's inhabitants. When he got his first chance to actually live underwater, he says the experience was magnificent. "I couldn't believe that when I was low on breathing air all I needed to do was to stand in the *Hydrolab* hatch and get a fresh tank." He remembers spending hours and hours in the water and feeling like one of the fish schooling around the habitat. With the advantage of time and the ability to observe fish in their natural environment, Bob saw enough in fifty hours to write several scientific notes on fish behavior.

Over the years, Bob and his team learned a tremendous amount about undersea habitats, to the great benefit of the scientific and diving community. They discovered that humidity inside the undersea station was a huge problem, at times reaching 100 percent, making it hard to breathe and to sleep. The high humidity also took a heavy toll on equipment and scientific instrumentation, and caused some serious skin and ear infections (both of which can still be a problem). Temperature controls and air-conditioning were among the improvements they made to *Hydrolab*. Bob recalls one important and rather unpleasant lesson—portable toilet facilities were not well suited to the early habitat configuration. During decompression, when the air pressure inside the habitat was decreased, the internal air inside the toilet's holding tank not only expanded, it literally exploded, splattering its contents all over the entry trunk of the habitat. In the more modern habitat, *Aquarius*, there are two options for going to the bathroom. There is an inside toilet with a curtain around it, but this offers little privacy and sometimes it gets clogged—resulting in a very dirty and unpleasant cleanup job. Therefore, just as became the norm for those living in *Hydrolab*, the other option is the ocean version of an outhouse and the method many aquanauts prefer. Some people refer to this as "becoming one with the sea" or "feeding the fish." However, there is a wee problem. Some species of fish consider human waste "tasty treats" and can be quite aggressive. On at least one occasion that I know of, an overly eager fish drew blood from a male aquanaut—from a very sensitive body part.

Early on, the idea of living underwater also attracted some notable visitors to *Hydrolab*, both in the Bahamas and when it was stationed in St. Croix. Bob fondly recollects a visit by the saturation-diving guru George Bond, along with thirty-five medical doctors. Famed submersible engineer Ed Link visited, inspecting the habitat for the Smithsonian Institution, and the late Peter Benchley, the author of *Jaws*, spent a night in *Hydrolab*, writing about his experience for a magazine. And when the prime minister of the Bahamas dove on *Hydrolab*, his bodyguards, who came dressed in suits and ties, became quite frantic at the surface when their charge disappeared into the sea. Bob says the bodyguards were a wreck by the time the prime minister surfaced. During the dive, John Perry (the original builder of *Hydrolab* and a well-known diver, engineer, and ocean advocate) scraped his leg on something and later remarked to the prime minister that he had bled for him. The prime minister replied good-naturedly, "But, aren't you glad it wasn't my blood." Other distinguished visitors included Wernher von Braun, the father of rocket science; the United Kingdom's Prince Charles; and the U.S. secretary of commerce Pete Peterson. When *Hydrolab* was in St. Croix, Senator Lowell Weicker, a fervent diver and strong ocean-science advocate, made several visits to the undersea laboratory.

After years of successful service, *Hydrolab* was eventually retired and put on display at the Smithsonian Institution. The habitat was later moved to NOAA's museum in Silver Spring, Maryland, where it resides today.

TODAY'S UNDERSEA RESEARCH HABITAT — *AQUARIUS*

There is currently only one operating undersea research station in the world—*Aquarius*. As the successor to *Hydrolab*, *Aquarius* is larger, more comfortable to live in, and technically more sophisticated. Its base sits at a depth of about 60 feet (18 m) in a sandy expanse near Conch Reef within the Florida Keys National Marine Sanctuary. The *Aquarius* is owned by NOAA and operated by the University of North Carolina at Wilmington. The main living section is cylindrical, about 43 feet long by 9 feet in diameter (13 m by 3 m), and is constructed of 2-inch- (5-cm)-thick steel, weighing 80 tons. At the seafloor, *Aquarius* is bolted to a four-legged platform that weighs another 100 tons. Its round windows or viewports are made of 2-inch- (5-cm)-thick acrylic plastic, and attached at its base are large air and water tanks. Power and communications are provided through cables that are connected to the large life-support buoy, or LSB, which floats above the *Aquarius* and hosts two generators, two compressors for air, and a wireless bridge for communications. The staff hopes to eventually move all of the power and air supplies to the seafloor and replace the LSB with a smaller buoy and fixed communication tower. Missions typically last one to two weeks, with an aquanaut team of four scientists and two habitat technicians from the staff as well as a surface-based crew.

The undersea staging area into the *Aquarius* is the wet porch. Because the atmospheric pressure inside *Aquarius* is kept at exactly the same pressure as the surrounding water, the hatchway or moon pool into the wet porch is left open (no water flows in). It is an amazing thing to stand in the moon pool with your upper body in the air environment of the habitat and your feet in the ocean, some 50 feet (15 m) down. Aquanauts shower and wash their gear in the wet porch before drying off and entering the interior portion of the research station, where air-conditioning keeps it dry and cool.

Inside *Aquarius* are redundant monitoring sensors and controls for the life-support system (air, pressure, carbon dioxide, temperature, etc.) and communications. There are areas for working on gear and experiments, along with sinks for washing, a tiny kitchen (sink, microwave, and cooler), a table for dining and working, and a bunkroom with two tiers of three bunks each. Aquanauts typically spend their days diving outside of the habitat as much as possible and eating, sleeping, working up data, or repairing equipment while inside. Within *Aquarius*, scientists now have access to the Internet and can post or broadcast mission updates and communicate with colleagues, friends, and family via e-mail.

An exciting advance in technology at *Aquarius* is that aquanauts can fill their scuba tanks at the habitat or at undersea way stations out on the reef. The way stations are small hut- or gazebo-like structures made of steel, which are partially filled with high-pressure air. Two divers can stand inside with full gear and be in air from about the waist up. An extra hose on their specially configured double scuba tanks can be hooked up to a valve inside the station and then, with just the simple turn of a knob, the air in their tanks can be quickly refilled. On one mission we dove for hours with a maximum depth of about 110 feet (33 m) by repeatedly filling our tanks at a way station on the reef. We also stored water and small snacks inside to refuel our bodies while we refilled our air. It was a fantastic experience that allowed us to accomplish coral and fish surveys that would have taken many days if we had been working from the surface. There is also an open two-way microphone in each station, so that the staff in the habitat can keep track of and communicate with the divers. At *Hydrolab* there were less advanced, small "bubble" stations where a diver could pop up in a lens of high-pressure air under an acrylic dome to confer with his or her buddy. These also had open microphones, and one beloved story is about a couple of recreational divers who stumbled upon a bubble dome and went inside to investigate. They were undoubtedly shocked when a loud, godlike voice issued forth, saying, "Don't touch that." The staff onshore happened to hear the divers enter the bubble and couldn't help but have a little fun with them.

While living underwater can be fun, exhausting, exhilarating, and a true adventure, the real goal is time underwater for science. As a unique field station, *Aquarius* provides state-of-the-art technology to do both applied and basic ocean research. Aquanauts have established long-term coral reef monitoring stations where they can do repeat, time-consuming surveys. Researchers can run twenty-four-hour tests of instrumentation and conduct technically difficult experiments in situ. The latter is especially advantageous because many marine organisms don't do well in a laboratory setting, and results in an isolated, artificial environment may not apply to the more complex, real world undersea. Chris Marten, Niels Lindquist, and Jeremy Weisz from the University of North Carolina at Chapel Hill used *Aquarius* to show that when measured in the field, one species of sponge pumps ten times more water through its system than suggested by previous lab-based experiments. Their experiments confirmed that sponges act as efficient and energetic filters on the coral reef; recent results indicate that they may also play a surprisingly important role in the cycling of nitrogen within the ecosystem. And they are using *Aquarius* to test and develop what may be one of the world's first underwater mass spectrometers—a high-tech instrument to measure a suite of chemicals or substances in the water.

In total, over ninety missions have been done in *Aquarius* since 1993. While living underwater, scientists have studied coral biology and diseases that remain poorly understood. They have drilled through the reef to look at its history of growth and done groundbreaking research on the interactions of herbivores. Using *Aquarius*, James Leichter of Scripps Institution of Oceanography discovered frequent upwelling events on the reef that are caused by tidal forcing. He recorded periodic surges of cold water with nutrient concentrations some ten to one hundred times the ambient conditions. His results are critical as we try to better understand how Florida's coral reef ecosystems function and how they are changing over time.

Study of the coral reefs surrounding *Aquarius* is particularly beneficial because the undersea station is in a specially designated marine protected area in the Florida Keys National Marine Sanctuary. All commercial and recreational fishing, diving, and collecting are prohibited in the area surrounding the habitat. Researchers can set up long-term surveys or test equipment without disruption by recreational divers, boaters, or fishermen, and while monitoring change on the reef, they can also help to assess how no-fishing zones affect a reef's dynamics. Early results from monitoring at *Aquarius* suggest that corals at deeper depths on the reef appear healthier than those in shallow water where increasing incidents of disease, bleaching, and overgrowth by algae have occurred.

Aquarius has also proven an effective tool for outreach and education. In cyberspace, students and teachers can now follow missions as they happen and watch aquanauts and the reef via real-time cameras. NASA is using the undersea research station as a means to train astronauts who will go to the International Space Station or the moon to set up a lunar base. In the sandy rubble around *Aquarius*, astronauts are testing robotic vehicles and the weighting and balance for their space suits, and practicing tasks that will be needed to establish a station on the moon. The U.S. Navy also uses *Aquarius* to train divers in saturation diving and to develop undersea technology.

UNDERSEA OBSTACLES

Along with the advantage of time underwater, *Aquarius* also presents its own set of challenges—not the least of which is keeping everything and everybody safe, operating smoothly and on schedule.

Scientists working in *Aquarius* face many of the same problems as other field-going researchers, such as equipment and experiment failures, but just diving for hours on end, even in the subtropics, can be problematic. On this issue Steve Gittings notes, "Spending nine hours in the water in a single day can make one miserably COLD." He remembers one April mission in which a part of the required postdive routine became a steaming cup of hot chocolate, and it never tasted so good. Of course, there is another way to warm up while diving for hours in a wet suit, especially after drinking lots of hot chocolate or coffee. Steve's advice is that it's best to enjoy the warmth, but not to think about where it is coming from, when you pee in your wet suit. Although the aquanauts' wet suits are washed each day, they do take on a rather pungent smell. This may also have something to do with why some aquanauts are temporarily plagued by body rashes—not-so-fondly nicknamed the creeping crud or the fungus among us. On his experience and the "comforts" of living underwater, Steve is steadfast: "I wouldn't have traded it for the world!"

THE MARTINI EFFECT

One of the more interesting, some might say amusing, problems faced by people living underwater is a phenomenon known as nitrogen narcosis—also called the martini effect. Much of the excess gas absorbed in a person's body while living underwater is nitrogen—think of nitrous oxide, aka laughing gas. After several days undersea, aquanauts can get a little silly. On one mission after several long days of diving, we watched one of the Austin Powers movies on a DVD. Under the effects of nitrogen narcosis, the movie was not only amusing—it was downright hilarious, inducing side-splitting, tear-producing laughter. One scientist, who shall go unnamed, later took to his top bunk and emulated Dr. Evil's lasers with his reading lights—the martini effect for sure.

Prior to one mission, I spoke with a producer at the *Today Show* to discuss plans for a live interview from *Aquarius*. With nitrogen narcosis in mind, I asked if I could get the questions that might be posed on air. The producer seemed taken aback by my request and informed me that such questions are never given out in advance. I then inquired if he was a scuba diver or had ever heard of nitrogen narcosis. After I explained the infamous martini effect, he gave me a list of potential questions, probably wondering if the upcoming interview, live, was such a good idea. I had a lovely and coherent chat with Matt Lauer while living in *Aquarius*, though later that day when we did a live Web chat via phone with MSNBC, there was a moment of hilarity when one member of our team was asked about sponges. His reply, to the amusement of the rest of us, was something along the lines of sponges spend their days sucking and blowing, after which he quickly rephrased it to say that sponges constantly pump water through their osculum and pore openings, but of course under the influence of nitrogen narcosis, we were already rolling with laughter. For the most part, nitrogen narcosis does not disrupt work underwater, as aquanauts are able to stay focused, but it does add a bit of humor during off-hours undersea. For safety, however, most of the dive planning is done during the shore-based week of training, and a team at the surface monitors the divers and life support 24/7. Of course, I cannot tell all of the nitrogen narcosis stories, because, as they say, some of what happens undersea stays undersea. But there have been some pretty funny practical jokes at *Aquarius*, several of which involved fluorescein dye, which becomes bright yellow when released into the water. Well, they were very funny at the time, or maybe it was just the effects of the nitrogen . . .

While not a gourmet's delight, the food in *Aquarius* can provide additional moments of humor. In the habitat's high-pressure atmosphere, air gets compressed. Bags of chips become flattened bags of bits, and packages of chocolate candy are transformed into abstract art—like bumpy balls. And jars (or camera lens covers) must be unsealed before being brought down or the small layer of air under the lid becomes a human-proof vacuum seal. Reconstituted freeze-dried meals have long been the norm while living in *Aquarius*, though every so often the aquanauts get a special delivery. One time we were treated to a diver-delivered pizza. Another time it was a kind gesture in the form of a homemade lemon meringue pie. But alas, because meringue is mostly air, it became a thin layer of white slime over what, in the land of bland undersea, was an unappetizing pie of yellow goo. On another occasion my colleagues asked for and were provided with sushi. Personally, I just couldn't bring myself to feast on raw fish with the meal's cousins looking in through the viewports.

Camaraderie with one's colleagues, trust, and a good sense of humor are essential to successful missions undersea. In fieldwork in general, humor is invaluable: it helps us to confront and deal with dangerous or frustrating situations and to relieve the stress from physical or psychological demands.

HURRICANES

When storms pay a visit to *Aquarius*, they are once again a demonstration of nature's raw power. For Craig Cooper, the operations director at *Aquarius,* storms, especially hurricanes, are a major problem and have forced his team to cancel or abort missions as well as recover aquanauts under "highly unfavorable" conditions. He is intensely proud of the fact that no scientists or any of the staff have been injured due to storms, although the habitat has taken a few hard knocks. The staff at *Aquarius* constantly monitors severe weather, typically tracking storms well before they reach the Florida Keys. They usually have time to prepare, except when storms develop unusually fast or move in more quickly than expected.

Craig remembers well the Groundhog Day storm of 1998, which hit unexpectedly with seventy-mile-an-hour winds. The massive life-support buoy broke loose from its mooring system and was driven onshore by strong wind and high waves. Fortunately, the incident happened between missions, and nobody was living in the habitat at the time. In 1994, however, when Tropical Storm Gordon took an abrupt jog from its forecasted track and struck head-on, it was midmission. They needed to get the aquanauts out of *Aquarius* as quickly as possible: this meant starting decompression almost immediately. During decompression the inside chamber of *Aquarius* is sealed, and over the next seventeen hours the pressure within is slowly decreased to one atmosphere, the same as at the surface. High seas during Gordon damaged the generators for *Aquarius*, so the aquanauts had to go through decompression under emergency battery power. They had light, communications, and, most importantly, scrubbing of carbon dioxide from the atmosphere within the habitat, but no air-conditioning. It was a long, hot seventeen hours below the sea. For Craig and the others at the surface, it was no party either; they had only flashlights during their supervision and control of decompression. Add to that a serious adrenaline rush during the night when their 28-foot (8-m) safety boat broke free of its mooring and nearly capsized. After saving the boat and completing decompression, the surface crew had to recover the aquanauts in 15- to 18-foot (5- to 6-m) seas. Two aquanauts at a time were brought up along an ascent line from *Aquarius* and quickly put aboard the boat, which stayed under power the entire time to prevent capsizing as large waves hit it broadside. To reach shore, they had to then navigate the high storm waves breaking over nearby Conch Reef. Craig recalls one of the boats temporarily disappearing from view in the 15-foot (5-m) waves as white water crashed over the reef. They all made it to shore without injury and returned to the habitat several days later to see what the storm had done. Scuba tanks that had been stored 60 feet (18 m) below the surface were stripped of gear, filled with seawater, and, most surprisingly, exhibited burnlike marks. The only plausible explanation—a lightning strike on the surface.

In 1998 Hurricane George also passed 80 miles (128 km) to the south of *Aquarius*. Instruments on site recorded 28-foot (9-m) waves before being torn from the bottom. Inspection of the habitat after the storm revealed that it had come frighteningly close to being destroyed. *Aquarius* had been severely shaken, so much so that a joint in one of the base-plate legs was broken and a 4-inch- (10-cm)-diameter pin of stainless steel was seriously bent. Two 8,000-pound weights on the wet porch were moved and close to falling off the structure. Two of the gazebo-like way stations outside of the habitat were destroyed, battery pods were stripped off the top deck, and one of the steel way stations was "ripped in two and left like a turtle on its back." According to Craig, it took several months to fabricate replacements for the equipment and to restabilize *Aquarius*. A month later Hurricane Mitch blew through, leaving the sole surviving way station a tangled mix of steel and fiberglass. Since then the habitat has been better anchored to the seabed, and ballast weights have been attached in a way that don't allow for movement. In addition, when time allows, storm preparations now call for the recovery of all way stations and battery pods.

In 2005 the sheer power of a hurricane, both at the surface and underwater, was once again showcased at *Aquarius* and left the staff wondering if it was the end of the program. Hurricane Rita broke two of the habitat's four seabed anchors and moved the wet-porch end of the habitat 12 feet (4 m) across the bottom. The habitat itself was unscathed, but repair of the damage to the structure required six weeks of solid diving and the help of U.S. Navy divers. Two days after all the repairs were done, Hurricane Wilma hit the Keys, but this time there was no damage, and just two weeks later missions resumed. Craig concludes, "Over the past thirteen years we've seen enough storms at the *Aquarius* to have gained a true respect for what they can do."

UNDERSEA WONDERS 24/7

The time availed by living underwater and working long hours on a reef provides ample opportunity to also encounter the wonders of nature undersea. Once, while doing a survey of reef fish, we looked into the shadows under a ledge and saw two very large eyes staring back at us. Soon a gentle giant emerged. It was a goliath grouper so large that one gulp of its enormous mouth could have sucked us in—literally. It swam lazily with us for several minutes, and a video of the encounter shows it to be well over diver size. The experience made us wonder how the enormous creature had stayed so well concealed for so long and if there were other behemoths lurking about unseen. On another day, when the water was incredibly blue and clear, we came upon a school of hundreds, if not thousands, of small silvery fish. As we moved through the school, the fish parted in silent, graceful synchrony. The movements of the fish were amazingly well choreographed and timed—a true natural wonder that is poorly understood.

One of my favorite times in *Aquarius* is the early morning, when my fellow aquanauts are still asleep. It is the perfect way to watch as morning breaks in the ocean and on the coral reef. As the surrounding sea goes from black to navy to a light royal blue, shafts of yellow light from the rising sun pierce through the water and flicker across the seafloor. Schools of fish return from their nocturnal forays, with some species arriving at the same place at the same time every morning, like a line of workers exiting a factory after the night shift. As the ocean turns a lighter shade of blue, other fish begin to emerge from their nighttime shelters, and soon the hustle and bustle of the daytime reef begins once again. And as I watched, so too did the fish. Every day several schoolmasters—a silver fish with yellow fins and tail, large eyes, and puffy down-turned lips—would swim by, stop, come back, and look in through the viewport. Just who was watching whom?

For scientists, the ability to live underwater can be the opportunity of a lifetime, enabling time so rare and so valuable in the ocean that it can significantly enhance their research, alter their careers, or forever change their perspective on the sea. While giving talks about living underwater at a wide variety of venues, I have also seen firsthand the power the *Aquarius* program has to excite and engage people of all ages in learning about the ocean, technology, and science. People often ask me if we will be living undersea anytime soon. We are nowhere near the vision of years ago in which humans had permanent homes amid gardens of coral. *Aquarius* now stands as the world's lone undersea research habitat and along with its shore-based field facility is struggling for survival because of reductions in funding. Its precarious financial status exemplifies what is happening to field stations around the world.

Field stations provide scientists and students access to the sea. They supply essential support for scientific research, exploration activities, and invaluable experiential learning opportunities. Field facilities, however, don't come cheap, they are difficult to run, and the rewards are rarely immediately apparent, quantifiable, or financial. As inhabitants of a planet whose surface is nearly three-quarters ocean and as a society that heavily relies on and impacts the sea, can we really afford not to invest in programs that help us to better understand the ocean or that can engage students and the public in learning about the sea and science?

Hydrolab. Photo courtesy of Dale Anderson.

The voracious grazing and prodigiously pooping parrotfish that contribute to nutrient recycling and sand production on coral reefs. Photo courtesy of Mike Schmale.

(*This page and the previous page*) *Aquarius* undersea research habitat off Key Largo, Florida. Photos courtesy of Dale Anderson.

Gazebo undersea way station at *Aquarius*. Photo courtesy of DJ Roller, Liquidpictures.

Packing the pot for the beginning of a mission in *Aquarius*. Sets of double scuba tanks for aquanauts lined up along the sides of the boat. Photo Ellen Prager.

Undersea vista in St. Croix, early 1980s. Photo Ellen Prager.

Pilot Rich Slater preparing for a dive in the submersible *Delta*. Photo Ellen Prager.

Top, the deep-diving submersible *Alvin* lifted from the sea. *Bottom*, photo taken from *Alvin*: a deep-sea chimney spewing black smokelike mineral-rich fluid at an active hydrothermal vent. Photos courtesy of Woods Hole Oceanographic Institution.

Otherwordly meadow of deep-sea crinoids. Photo courtesy of Chuck Messing.

Diver surrounded by cobia inside an "Aquapod" submerged cage off Culebra Island, Puerto Rico. Photo courtesy of Daniel Benetti.

7   *Going Deeper*

THE DEEP OCEAN IS ONE OF THE MOST EX-treme environments on the planet. It is cold and dark, and the pressure is enormous. It is also mostly unexplored. For scientists who study the deep sea, it is an exciting yet dangerous challenge whose rewards come from unveiling a world in which the creatures and their surroundings have rarely, if ever, been seen and that are far different from those we are familiar with. Just to get there is a technological feat.

In 1960 Captain Don Walsh and his fellow explorer Jacques Piccard became the only humans to have ever journeyed to the very deepest part of the ocean, the Mariana Trench, over 35,000 feet (10,900 m) below the surface. It took them five hours of descending in the *Trieste*, an early submersible or bathyscaphe, for a brief twenty-minute stay at the bottom. When Don speaks of the experience, he points out that key to their mission was making it a two-way trip: "After all, it's the return leg that really counts." Surprisingly, today, over four decades later, there are no manned or unmanned undersea vehicles capable of reaching the depths probed by Walsh and Piccard.

HIGH-TECH TOOLS

From manned, deep-diving submersibles to undersea robots, the benefits of technology shine in the deep ocean. Unlike shallow-water ocean environments, we cannot just throw on some fins and jump in. People who want to explore the sea's deepwater realms must rely heavily on technology designed specifically for the task.

Deep-diving submersibles are built to traverse great depths of water, to withstand extremes in temperature and pressure, and to be compact enough to allow for transport, launch, and recovery from a support ship. They must also be highly maneuverable, while flexible enough in design so that they can be outfitted with a wide array of scientific instruments, such as sonar, cameras, lights, a manipulator arm, and sample holders. In the world today, there exists a small fleet of submersibles, the majority of which can access only the sea's shallow to midwater depths. Just a handful of the current manned undersea vehicles can journey below about 3,000 feet (900 m); these include Russia's *Mir I* and *II*, France's *Nautile*, Japan's *Shinkai*, and the United States' *Alvin*.

Researchers can also probe the ocean's depth, without actually having to go there, by deploying remotely or autonomously operated vehicles. Remotely operated vehicles (ROVs) are underwater robots that are tethered by a cable to a support ship, from which scientists can guide their movements and sampling. Autonomous underwater vehicles (AUVs) are free-swimming robots or gliders that can be programmed for specific tasks and deployed independently. They travel and sample as programmed and then return to a base of operations or ship. Both ROVs and AUVs are cost-effective and especially advantageous in highly dangerous or inaccessible environments, such as in the deep sea, under the ice, or within undersea caverns. However, even robots can run into trouble. In 2003 the deep-diving Japanese ROV *Kaiko* was lost during a typhoon when a cable connecting it to the surface snapped. In 2005 an AUV was deployed under the Antarctic ice but failed to return, and remains missing in action. In the deep sea, marine engineering is particularly critical, and it is at the heart of success in ocean science. One place where high-tech undersea tools have been showcased and are prov-

ing incredibly efficient and productive in deep-sea exploration is
Monterey Canyon, off the coast of California.

A STRANGE DEEPWATER WORLD

Marine biologist George Matsumoto and a team of researchers at
the Monterey Bay Aquarium Research Institute (MBARI) regularly
cruise off California's coast to explore the depths of Monterey
Canyon as well as nearby locations. From their ships they deploy
ROVs that can go thousands of feet below the surface and are outfitted with a high-definition video camera. George has been studying
deep-sea animal life for almost twenty years, yet he continues to be
mystified and fascinated by the strange creatures they consistently
encounter. Some of the organisms his team finds are completely
new to science, animals that have never before been seen by human
eyes. One such creature is an odd-looking, softball-sized, translucent
jellyfish with a bell and feeding arms that are covered with wartlike
bumps. Though it has no tentacles, the jellyfish's bumps are actually
clusters of stinging cells that are used to capture prey. MBARI
scientists discovered "Bumpy" in the dark depths between 500 and
1,800 feet (150 and 550 m) and named it *Stellamedusa ventana*. The
first half of its name (its genus), *Stellamedusa,* was chosen to reflect
the animal's translucent blue-white color and trailing arms, which
reminded the researchers of a slow-moving meteor or shooting
star. The species name, *ventana,* refers to MBARI's ROV *Ventana,*
which first recorded the jellyfish on video in 1990. For George,
however, just seeing such animals isn't enough. He wants to know
more: "What are they doing? What are they eating, and who is
eating them?" Yet if viewing deep-sea animals is difficult, learning
more about their daily activities and behavior verges on the impossible—for now.

Monterey Canyon is one of the largest submarine canyons in
the United States and a deep-sea environment that is relatively close
to shore. With a world-class research institution (MBARI) nearby

that specializes in deep-sea exploration, the region has become a hotbed for the discovery of mid- and deepwater species. It is also part of the Monterey Bay National Marine Sanctuary, which includes the nation's largest kelp forest and one of the most diverse marine ecosystems in the world. Within the sanctuary, researchers have identified 33 species of marine mammals, 94 species of seabirds, 345 species of fishes, and numerous invertebrates and plants. Add to that the new deepwater species that are regularly being discovered in the canyon's depths. The sanctuary's proximity to shore also offers the public a wonderful opportunity to experience "the field" through local diving, boating, and kayak excursions or, for the less adventurous, at the Monterey Bay Aquarium.

The use of both ROVs and AUVs has been a true revolution for scientists studying and exploring the deep sea, but there is still something that even the highest of high-tech cannot yet replicate—the power of firsthand experience.

FIRSTHAND EXPERIENCE

Renowned deep-sea biologist Shirley Pomponi has spent years diving in submersibles. Her research focuses principally on the search for pharmaceuticals derived from marine organisms. With the greatest biodiversity on the planet, exceeding even that of the rain forest, the sea is truly an untapped storehouse of potential disease-fighting drugs. Shirley is particularly interested in the chemicals produced by benthic creatures in the deep sea or on coral reefs. Organisms that live attached to the seafloor have no physical means to escape or avoid predators, so many have evolved chemical defenses.

Shirley's success and passion for deep-sea exploration exemplifies both the tangible and intangible benefits of manned exploration and fieldwork in the sea: "I love being at sea and diving in a sub! There's so much yet to explore and discover." On almost every dive, she sees something new or different. Because ROVs and AUVs use brand-new technology, Shirley believes that people consider them "sexy" or in vogue. "Sometimes I feel like I'm part of a dying breed," she laments, believing that unless you've actually "been there," it's hard to explain how much more the human brain can process as compared to even the best high-definition video, multibeam sonar, and remote sensor. And when decisions need to be made about where to go or what to sample, they can happen right there and then because in a submersible there are only a few people present.

Shirley's work and her commitment to oceangoing fieldwork are even more impressive given that she is prone to seasickness. The only saving grace is that if the seas are rough enough for her to start feeling queasy, it is usually too dangerous to launch and safely recover a submersible. On those occasions she retires to her cabin, drinks ginger ale, eats crackers, and lies on her stomach till the feeling passes or the seas die down. Having tried just about every remedy available and because motion-sickness medications make her feel even worse than the seasickness itself, she is now trying a bracelet that emits a slight electric shock every few minutes. Shirley theorizes that she is so focused on when the next "shock" is going to happen (it sends a tingle halfway up the inside of your arm), that she doesn't think about being seasick.

The power of experiencing the deep sea firsthand is also plainly evident when talking about submersible dives and deep-ocean crinoids with expert Chuck Messing. He describes a specific region rich with stalked crinoids, or sea lilies, at a depth of some 1,300 feet (400 m) as an otherworldly meadow. The crown or flowerlike top of each crinoid was oriented into the ocean's flow to facilitate filter feeding, reminding Chuck of a field of radar dishes all turned in the same direction. Crinoids are some of the ocean's oldest living organisms, having evolved over 400 million years ago. When noted paleoecologist Adolf Seilacher joined Chuck on a submersible expedition and saw the deep-sea meadow of crinoids firsthand, he likened it to a scene from the movie *Jurassic Park*, only in this case it was "Jurassic Aquarium." Chuck also tells of when a sub accidentally decapitated a crinoid and the scientist aboard got a big surprise—it crawled away. At the time, the species of stalked crinoid involved in the incident was believed to live its entire life fixed in place. The decapitation incident and subsequent observations by a number of scientists led to the conclusion that these stalked crinoids are actually mobile and can indeed crawl about the seafloor.

DOWN TO *TITANIC*

In 2000 surgeon Michael Manyak took part in an archaeological expedition to the RMS *Titanic* as the team's medical director. During the expedition he was given the opportunity to go down in one of Russia's deep-diving *Mir* submersibles. His journey into the deep sea began with a memorable briefing aboard the sub's mother ship, the *Keldysh*. He and the other novices aboard were told that at the depth of the *Titanic* (13,000 feet or 3,962 m) even a small leak in the submersible would create a jet of water so powerful that it would cut a person in two, but not to worry. Way before that happened, the sub would implode with such force that they would be instantly incinerated, even before being crushed. Mike was later glad to learn that the most experienced sub pilot would be at the helm for his dive. The night before his descent into the ocean's depths, he felt a mix of anticipation and excitement similar to what he feels just before performing a complicated major surgery. He spent a lot of time that night reviewing and re-reviewing a checklist of responsibilities.

The next morning on the deck of the *Keldysh*, Mike climbed inside the *Mir* submersible. His view of the world instantly became limited to a small 8-inch- (20-cm)-thick acrylic viewport nestled within the sub's 1.5-inch (3.8-cm) reinforced titanium hull. In a well-coordinated and rehearsed process, the crew used a crane to lift the sub off the deck and lower it into the water. Once disconnected from the mother ship, the submersible began a slow spiraling descent, sinking at about 82 feet (25 m) per minute. On the way down, the pilot repeatedly checked the lights, oxygen, and carbon dioxide levels, as well as the communication system. Down to about 500 feet (150 m) the visibility outside the viewports was good, but those aboard the sub saw little marine life. As they descended farther, the pilot turned on the vehicle's outside lights. The water was surprisingly clear, and Mike could see an amazing assortment of marine life, "everything from tiny unidentified creatures paddling frantically about to bright red shrimp and intricate jellyfish, one with a bright red internal globe, like a Christmas ornament." A large rattail fish curiously approached the sub, while other ghostlike creatures floated by. After descending for about two and a half hours, they reached the sandy seafloor, where it was exceptionally calm and there were just a few deep-sea corals scattered about.

Sonar revealed that the submersible was some 3,200 feet (1,000 m) from the *Titanic*. As they moved closer, Mike began to see debris, and then suddenly the giant bow of the ship came into view. He remembers it well, a ghostly, magnificent sight, even more awesome than he had imagined. They inspected the huge tear in the *Titanic*'s midsection that sheared the ship's stern from its bow, which now lay nearly 2,000 feet (600 m) apart. They also viewed the damage inflicted by the infamous iceberg as it scraped across the hull. The sub then cruised over the large debris field that surrounds the ship's stern. Scattered across the seafloor were dishes with the White Star Line logo, pieces of furniture, personal items, chandeliers, portholes, and candelabras. As the sub cruised over these reminders of everyday life, Mike recalls, "The tragedy of the *Titanic* was made ever more real and horrific."

During the dive they recovered a total of seventeen artifacts, the most significant of which was the telegraph that connected the bridge to the engine room. When the iceberg was sighted on *Titanic*, a lever on this telegraph would have been pushed in an attempt to quickly change course and speed. Mike was impressed by the extreme care taken in the recovery of the telegraph and all of the artifacts. Each item was carefully logged at discovery, videotaped in situ, and given to the curator and chief marine archaeologist on the *Keldysh* for identification and preservation. Nowhere did they find human remains, which Mike hopes will dispel the myth that biological material is still present on the site.

They spent a total of six hours at the seafloor during Mike's dive to the *Titanic*, after which the sub began a slow ascent. It took three hours to reach the surface, and after a bit of bobbing on the swells of the North Atlantic, the sub was hoisted back aboard the *Keldysh*. To this day Mike still feels the exhilaration of clambering out of the *Mir*'s hatch to the cheers of the crew and remains utterly grateful that he got to experience the deep sea in a way that few people ever will.

SCIENCE IS EXCITING

Even for researchers who study the deep ocean, the experience in a submersible can be astonishing, so much so that it can be hard to stay focused on science. In these instances enthusiasm may turn objective scientific narrative into the layperson's words of excitement—*ooh, ah, oh my god,* and *holy shit* are a few of the descriptors that have been uttered under such circumstances, some even recorded for posterity. Deep-sea geologist Susan Humphris of Woods Hole Oceanographic Institution vividly remembers her first dive in the submersible *Alvin* to observe hydrothermal vents. After descending for two hours in the *Alvin*, she got her very first look at an active undersea hot spring. "It was an incredible chimney covered with billions of shrimp with black smokelike fluid pouring out of it." Susan was so astonished by the view that her natural reaction was to exclaim in not-so-scientific terms. She then remembered that her verbal description was being recorded as part of the submersible dive and the warning her PhD adviser had given her. He said that on almost everyone's first dive to the vents they ended up uttering expletives of shock and awe rather than providing a useful scientific description for the record. Susan recalls that it took great self-control that day to appropriately describe the amazing deep-sea vents and organisms she saw. Even after years of studying hydrothermal vents, she still fondly remembers that very first dive, the sense of discovery it inspired, and how self-discipline was needed so that her enthusiasm did not overshadow her duties as a researcher. To this day that same sense of discovery and excitement is obvious when talking to Susan about fieldwork in the deep sea and especially about her work on the ocean's amazing undersea hot springs.

Whether shallow or deep, in the *Mir* or *Alvin*, collecting artifacts from the *Titanic*, exploring hydrothermal vents, or simply viewing the creatures within the ocean's depths, all of my colleagues agree: each and every dive in a submersible is an exciting voyage of discovery, both personal and scientific. With so little of the deep ocean having been explored, there is always a strong possibility of finding something new, different, or unexpected. Since the discovery of hydrothermal vents in 1977, more than 600 new species of associated deep-sea life have been found. One group of scientists returned to a hydrothermal vent site on the East Pacific Rise to retrieve their equipment and data only to find that their research area had been paved over by glassy, black basalt during a recent undersea volcanic eruption. Not so great for their equipment, but a tremendous, serendipitous opportunity to learn more about underwater volcanism and the creation of new oceanic crust, and to establish a site to monitor the potential regrowth and colonization of vent-associated organisms.

CHALLENGES DOWN DEEP

When I asked my colleagues who explore the deep sea about the most difficult obstacles they face in fieldwork, their answers echoed those of scientists working in shallow water—limited access, time, funding, the angst of proposal writing, and bureaucracy. For the most part they don't consider the discomforts, danger, or inconveniences of working in the ocean's depths as problems. Sitting in a small submersible for hours on end—cramped, cold, and cut off from the world above—is just part of the job. It takes a backseat to the amazing adventure and science of going into the deep sea.

Submersibles can, however, get stuck, snagged, or wedged in while deep in the ocean. They are therefore made to be highly maneuverable and are outfitted with emergency power, ballast release, and life-support systems. Sub pilots and the marine engineers that work on undersea vehicles are highly trained and diligent about safety, so serious accidents are extremely rare. The problems that scientists face on a more regular basis while in submersibles tend to be more frustrating than life-threatening. Take dripping water, for example. As a submersible goes from the relative warmth at the surface to the cold at depth, water condenses on its interior surfaces. This results in dripping water that can plague a scientist attempting to log a sample location or write a description of what is being seen. In addition to being an annoyance, dripping water is somewhat unsettling for obvious reasons. When Bob Halley worked in the small *Delta* submersible at 1,100 feet (340 m), water squirted in through the socket of a manually operated manipulator claw. Besides feeling like small stinging needles on his skin, it just seemed *wrong*.

For Susan Humphris, the limited view from the *Alvin*'s small windows is one of the most frustrating problems when exploring the deep sea. Since the two observers and pilot all look out of different viewports, they each have a unique view of the surrounding ocean or seafloor. She finds this especially annoying when trying to explain to the pilot where to stop to film a chimney spouting black, mineral-rich fluids or to collect a sample for later analysis in the lab. Susan solves the problem by crawling over the pilot to get his or her perspective and then trying to direct them to exactly the right spot.

In Harbor Branch Oceanographic Institution's Johnson Sea-Link submersibles, an observer and pilot sit within a 5-inch- (12.7-cm)-

thick acrylic sphere at the front of the vehicle. Unlike the *Alvin*, these submersibles allow for an awe-inspiring panoramic view; however, they can go down only to a maximum depth of 3,000 feet (900 m). Shirley Pomponi notes a different and rather surprising problem they have had in the Johnson Sea-Link subs. Three times now they have been attacked—by swordfish. She suspects that the swordfish were startled by the sub's lights and reacted instinctively with aggression. One swordfish rammed the sub so hard that part of its bill broke off and became lodged in the hatch entrance to the sub's acrylic sphere. They were at about 600 feet (180 m) off Key West at the time. Two days later as they dove in an offshore sinkhole, about 40 miles (64 km) north of Key West at a depth of about 1,800 feet (550 m), they discovered a dying swordfish with its bill broken off. Shirley recounts, "I didn't eat swordfish for at least six months . . . bad juju."

One of the basic challenges scientists face while working in the deep sea is deciding where to dive to find what they are looking for. Few areas in the deep ocean have been mapped in detail, and there are no road signs or travel guides to consult. Using the relatively coarse bathymetric data available, Shirley typically looks for steep rocky slopes that would provide a good substrate for benthic organisms to attach to. She says that when everything goes right on a submersible dive, they find what they are looking for, and if it is an interesting setting, the experience is simply thrilling. She remembers in particular one dive off the coast of San Salvador in the southeastern Bahamas. The trim and ballast of the sub were perfect, the visibility was remarkable, and the topography no less than spectacular. They were at a depth of approximately 2,500 feet (760 m), and the submersible seemed to be flying through the ocean. They traversed up, around, and within inches of a wall made up of rocky buttresses interlaced with deep furrows and caverns. And on the surrounding steep slopes were beautiful glass sponges, gorgonians (soft corals, such as sea whips and sea fans), and slit shells.

ALVIN AND THE NEXT-GENERATION SUB

The submersible *Alvin*, which is owned by the U.S. Navy and operated by the Woods Hole Oceanographic Institution, has long been the workhorse of the deep sea for researchers and explorers. Since 1964 the sub has logged over four thousand dives at depths averaging over 6,000 feet (1,800 m). The *Alvin* has been instrumental in maritime archaeology and in advancing our understanding of the deep sea from geology to chemistry, biology, and ecology. The submersible also assisted in the recovery of a lost hydrogen bomb in the Mediterranean in 1966 and was key to finding the RMS *Titanic* in 1986 by a team led by world-famous undersea explorer Robert Ballard.

Like other subs, the *Alvin* can be outfitted with a variety of diverse instruments for specific research needs, including a video camera, manipulator arm for collecting samples, sonar, sediment corers, and temperature sensors. On a typical dive, one pilot and two observers squeeze into *Alvin*'s spherical titanium hull for six hours. At the surface, a large support crew is needed to prepare, launch, and monitor the *Alvin* and operate the support ship, the R/V *Atlantis*. Trips to the deep sea are long and slow; they are also expensive. Operating *Alvin* and its support ship together can cost some $40,000 per day. The *Alvin* has also suffered its share of mishaps, the most serious of which occurred in 1968 while preparing for a dive off Cape Cod. While being lowered into the sea, the submersible broke loose from its steel cables and sank more than 5,000 feet (1,524 m) to the seafloor. Thankfully the pilot was able to exit with only minor injuries, and it went into the depths unoccupied. It took nearly a year before the submersible could be recovered.

After forty years of service, plans are to replace the *Alvin* with the next generation of U.S. submersible. The new undersea vehicle is being designed to explore more than 99 percent of the seafloor to depths of 21,320 feet (6,500 m). Along with the ability to dive deeper than *Alvin*, the new vehicle will be faster, more maneuverable

midwater, and able to carry an increased payload of scientific gear and samples. It will also have upgraded navigation and communication systems. The Deep Submergence Laboratory at Woods Hole Oceanographic Institution is building the new sub, and leading the project is marine engineer Bob Brown, the current manager of the *Alvin* group. According to Bob, the most exciting advance in the new submersible will be better viewing for the observers. He notes that a direct view of the seafloor has always been one of the strengths of human-occupied submersibles, but having the observer and pilot seeing a different field of view is problematic. In the new vehicle, their views will overlap. The observer will be able to directly see what the pilot is doing and sampling. Both pilot and scientist will also get the best view in the house—out the front of the vehicle.

The deep submergence group faces some unique challenges in building the next-generation submersible, the most difficult of which, says Bob, "is the design and fabrication of a personnel sphere with large and closely spaced viewports." Spheres thick enough to withstand the enormous pressure in the deep ocean are rarely designed or fabricated. He explains that they plan to use electron-beam welding to join the hemispheres and attach the viewport forgings, which will make it a first for deep submersible pressure hulls. Once completed, the new submersible, like *Alvin*, will face other obstacles in operations, not the least of which is maneuvering in geologically complex and dangerous terrain. At hydrothermal vents on a midocean ridge, there may be narrow valleys and towering mineral accretions, along with chimneys spewing superheated fluids capable of melting a sensor probe or damaging the sub. Such vents are no place for a sluggish sub.

The cost to build the next-generation submersible is currently estimated at about $22 million, with some of the funding already provided by the National Science Foundation and the Woods Hole Oceanographic Institution. Completion of the project will probably not happen before 2011, and even that remains questionable, because

not all of the needed funding is in place. Twenty two million dollars seems like a lot of money, but consider that one NASA space
shuttle costs $1.7 billion plus another $450 million per mission.
It will take a fraction of that to build and operate what will be
the nation's only deep-diving submersible—an investment many
people believe is essential if we are to fully explore, safely utilize,
and better understand the ocean.

8 *The Changing Sea*

ON EARTH CHANGE IS NATURAL AND VARI-
ability comes with time, but today there is growing
evidence that humans are causing an unwanted
evolution in the ocean. Astonishing quantities of marine debris
plague our coasts, washing ashore even in some of the world's most
remote locations. Thick blooms of algae, which can kill fish and
render the sea unlivable, are increasingly occurring along our coasts
and clogging our waterways. Across the globe, far fewer fish now
populate the ocean, corals are exhibiting signs of stress and dying,
and alarming ecological shifts are taking place. Whereas in the past
we viewed the sea as a vast, limitless entity whose resources could
be used without consequence, today the truth is inescapable. We are
changing the ocean and not for the better. For scientists who have
been going into the field for years, these changes are distressingly
apparent.

AN UNDERSEA GHOST TOWN

In 1962 graduate student Sonny Gruber was already becoming well
known for his way with sharks. He was part of a team that caught
and handled large lemon sharks for the filming of the James Bond
movie *Thunderball* in the Bahamas, and then took part in the big
underwater fight scene at the end of the film. Much of the action

was shot near a coral reef just off the exclusive enclave of Lyford Cay. At the time, Sonny was struck by the richness of the reef, especially because it was so close to land. Within swimming distance of the island, there were spectacular stands of branching coral and a wealth of beautiful reef fish.

In 1999, after a meeting in the Bahamas, Sonny took the opportunity to return to that same reef off Lyford Cay. He was appalled by what he saw. The entire reef was dead. There were few fish, and algae were growing over and on top of everything. The coral was still there, but it was all dead. It had become an underwater ghost town.

Sonny, like many of my colleagues, reports that this isn't the only place where he has seen such devastation underwater. Reefs around the world are exhibiting signs of decline, and many now host a mere fraction of their former fish and coral life. While the geological record shows that corals and reefs have always changed with time, what is being observed today is not the result of natural variability, but the unprecedented influence of humans. And it is not just coral reefs that are feeling the effects.

SEA SLUGS TAKE A HIT

Scientist Duane De Freese did his graduate research on the ecophysiology of sea slugs, which he describes as a diverse group of highly specialized and beautiful mollusks that don't get the respect or appreciation they deserve. Think of them as large, naked snails or shellfish without the shells. In the early 1980s sea slugs were abundant in many of Florida's nearshore habitats, but by 2000 many species had become so rare that they were locally considered close to extinction. Duane is saddened to think that we may be the first generation of marine scientists telling our students about species that we studied but that can no longer be observed in the field, because they simply don't exist.

The litany of problems in the ocean caused or exacerbated by humans is lengthy; along with coral reef decline and dramatically diminished marine populations, it includes rising incidences of harmful algal blooms; the loss of seagrass, kelp beds, coastal marshes, and mangroves; pollution from runoff and the atmosphere; the spread of nonnative or invasive species; and the effects of global warming. Field studies have helped to recognize and understand many of these problems; however, in most cases we still don't have an adequate baseline of data or the long-term, field-based monitoring that is needed to fully comprehend the changes taking place in the sea.

CHANGES UNSEEN

The consequences of human influence in the ocean can be stunningly apparent: take, for example, the collapse of the North Atlantic cod fishery or when pollution causes a massive fish kill or a vibrant green bloom of algae. Other changes in the sea are less visible, though no less alarming.

The concentration of carbon dioxide in the earth's atmosphere is increasing, in part at least, due to human influence. As a result, more heat is being trapped within the earth's atmosphere, and our climate is warming at an accelerated pace. We can now see some of the changes produced by this warming. Huge glaciers and ice sheets are melting or calving off into the sea. Polar bears are searching for disappearing food and ice floes. And people living in low-lying coastal areas, such as on the islands of the Pacific, are being forced from their homes as the sea rises. There are other changes taking place in the sea that are far less obvious.

As the planet warms, so do ocean temperatures. In response, some marine populations appear to be shifting their natural range. As carbon dioxide levels increase in the atmosphere, more is also being absorbed into the ocean, making it more acidic. We have yet to see the impact of this rising ocean acidity, but scientists are seriously concerned about its effects on biological processes in the sea and especially on marine organisms that create shells or skeletons of calcium carbonate, such as corals or the sea's tiny but abundant coccolithophores, the single-celled plants covered with microscopic plates of calcite that through photosynthesis contribute a significant amount of oxygen to our atmosphere. Scientific studies are needed, especially in the field, to monitor the chemistry of the sea and to investigate how rising ocean acidity will affect corals and other organisms that are highly susceptible to changes in the ocean's pH.

Other ecological shifts that are difficult to discern may also be occurring in the sea. Scientists have long known that when a dominant predator or grazer is removed from a specific marine environment, the ecosystem responds. Kelp forests provide an illustrative example. Within the kelp beds of the Pacific Northwest, the predators of herbivorous sea urchins, such as sea otters, fish, and lobsters, keep their populations in check. If one or more of these predators are removed, the sea urchins reproduce and graze uncontrolled, thereby turning lush kelp beds into barren bottoms of sand and rubble. In other ecosystems, the removal of plankton-filtering or herbivorous fish can lead to algal blooms in coastal waters or overgrowth on coral reefs. In 2007 Ransom Myers of Dalhousie University and his colleagues reported that large predatory shark populations along the U.S. east coast have been drastically reduced by overfishing, and its impacts are now cascading down the food web. Free from predation, one of the sharks' main prey species, the cownose ray, has multiplied in great abundance and feasts heartily on one of its favorite foods—the bay scallop. In 2004 a century-old bay scallop industry in North Carolina had to be shut down, because cownose rays had eaten most of the scallops. What other ecological shifts in the sea are taking place due to human influence? Are algae replacing corals on reefs worldwide? Will overfishing lead to coastal ecosystems where, as some science suggests, jellyfish will dominate over fish? Are increasing algal blooms and dead zones along our coasts going to be the norm, with productive and diverse fish populations the exception?

Humans depend on the ocean for food, health, safety, transport, recreation, and billions of dollars in economic revenue as well as millions of jobs. We are not separate from the sea, but are part of the ocean ecosystem, influencing and being influenced by it. The role of ocean science and fieldwork is now more important than ever, not only to explore and further understand the sea, but to help prevent further degradation and to find ways to sustain our use of its resources while preserving its long-term health as a productive and functioning ecosystem.

SURF'S UP

Fortunately, the ocean has proven to be resilient, so there is reason to hope, but time is limited. Our impacts on the sea can be curtailed and prevented, but it takes political will, public support, and sufficient understanding through ocean science to identify the problems and find solutions. A good example of the progress being made comes from the beaches of California. Many years ago, raw sewage was regularly released onto or near the shore, but this practice is no longer allowed. Treated wastes are now discharged far offshore where they are more readily diluted, mix with the surrounding seawater, and do less harm. Marine biologist Steve Weisberg with the Southern California Coastal Water Research Project Authority confirms that, as compared to forty or fifty years ago, California's beaches are much cleaner. There are now regulations about what can be discharged into the sea and where. There are still difficult issues to deal with, such as localized but poorly understood problems and nonpoint source pollution, which comes from runoff especially during wet periods, but Steve is optimistic that in the future, with the needed investment and political and public will, California's beaches will be even cleaner. The direct release of untreated sewage onto beaches and into coastal waters still occurs in many areas of the world.

SUSTAINABLE USE

Ocean science is also helping people find sustainable ways to use the sea's resources through activities such as responsible ecotourism. Erik

Zettler is the science coordinator for SEA. During a recent cruise as chief scientist, Erik along with the students and staff aboard the SSV *Corwith Cramer* collected some of the first oceanographic data from Samana Bay, the largest estuary in the Dominican Republic. While working in the bay, Erik collaborated with researchers from EcoMar, a research and education nonprofit organization based in Santo Domingo. They have been studying Samana Bay for several years and are working with local residents of a nearby village to help them transition from fishing to a more sustainable economy based on responsible whale watching. At the head of Samana Bay is a shallow bank and deep spot where humpback whales regularly congregate for the winter breeding season. Boat operators now go out with tourists to watch whales and while doing so help to collect much-needed data on their distribution. Local high school students and teachers are also being trained to make whale observations and often go along. Despite a language barrier, Erik and his students found it enlightening to discuss with local residents how they are coping with the changes taking place in their community. With fisheries in the area on the decline, they are optimistic that ecotourism will prove sustainable and improve the village's economy, now as well as in the future.

A RELATIVELY SAFE HAVEN

Two of the hottest topics in ocean science and policy these days are ecosystem-based management and marine protected areas. The two are related in that marine protected areas offer a mechanism to manage ocean ecosystems as a whole and take into account the influence of and use by humans. There are various kinds of marine protected areas, the most restrictive of which are no-take reserves. Scientific studies worldwide now show that protection from fishing in marine reserves leads to a rapid increase in the abundance and average size of previously exploited fish as well as an overall rise in the diversity of marine life in the area. Callum Roberts of Harvard

University and his colleagues have also found that the benefits of no-take zones can reach beyond their borders and enhance fish populations in adjacent waters. However, marine protected areas, especially no-take reserves, don't always go over well at first. In the Florida Keys, before the benefits of protection were realized, there was some very vocal opposition—it is a story that has been playing out at locations around the globe.

The Florida Keys hosts one of the largest areas of coral reefs in the world along with extensive seagrass beds, mangroves, and a wide diversity of marine life. By the late 1980s, however, even though there were already three national parks and two marine sanctuaries in the area, the ecosystem was at risk. Scientists had documented coral bleaching and disease, seagrass die-offs, and declines in the populations of reef fish. There were also concerns over declining water quality and proposals for oil drilling in the area. In 1989, when within an eighteen-day period three large ships ran aground and destroyed extensive areas of coral reef, Washington, DC, took note. On November 16, 1990, approximately 2,800 nautical square miles of state and federal waters in the Keys were designated as the Florida Keys National Marine Sanctuary.

Today, the Florida Keys National Marine Sanctuary receives widespread support, but that wasn't always the case. Initially, when public hearings were held to discuss the management plan for the area, there were mixed opinions and strong emotions about potential restrictions and the proposed creation of no-take zones. Environmental groups strongly supported the plan; however, fishermen and some dive operators were not so swayed. The fishing community felt alienated from the political process involved and was afraid of losing access. Dive operators had concerns about how their activities might be limited by any new regulations and zoning. Some antigovernment activists were even more vocal. Representatives from NOAA helping to develop and manage the sanctuary received death threats, had their tires slashed, and were hanged in effigy. Much of the original opposition was because the public didn't understand the sanctuary plan or its benefits.

In 2001 the Florida Keys National Marine Sanctuary was expanded to include the Tortugas Ecological Reserve, one of the largest no-take marine reserves in the world. For the most part, opinions in the Keys have changed dramatically, and some people are now calling for increased protection through further expansion of no-take zones. The experience in the Keys illustrates that good communication with and support of the local community is critical when establishing marine protected areas; it is also essential for long-term success.

The Florida Keys National Marine Sanctuary also provides an example of ocean zoning, a concept in use on Australia's Great Barrier Reef and along the coasts of Belgium, China, the Netherlands, and the United Kingdom. Ocean zoning is essentially like spatial planning done on land, *except* that in the sea it involves public property with multiple and conflicting uses and boundaries that are invisible and easy to cross. Within the Florida Keys National Marine Sanctuary, there are region-wide regulations that include no oil or gas drilling; no collecting of coral, tropical reef fish, or live rock; no discharging of trash or other pollutants; and no anchoring or vessel operations that will harm the seabed and organisms living there. There are also twenty-four smaller, fully protected zones, which are called ecological reserves, sanctuary preservation areas, or special-use areas. Restrictions in these zones, which encompass only about 6 percent of the sanctuary, are more stringent. Outside of the fully protected areas, responsible tourism through recreational fishing, diving, and boating is encouraged. Mooring buoys have been installed at popular dive sites along with improved signage to identify critical habitats and specially protected areas.

One undeniable and visibly successful aspect of the Florida Keys National Marine Sanctuary is the fish. I recently took the opportunity to dive with Sea Dwellers, a commercial operator in

Key Largo that brings divers and snorkelers out to the local reefs. The boat was full of tourists, and their positive experience illustrates the benefits the sanctuary has brought to the Keys. During just two short dives that day, an amazing array of marine life was seen. A school of large midnight-blue parrotfish swam by with big, fat groupers in its midst. Several divers were awed as graceful eagle rays glided by. I found several large spiny lobsters battling in an undersea crevasse and shiny copper sweepers schooling within a small cave. Other divers saw sharks, a large moray eel, and a sea turtle. And colorful reef fish were everywhere, from the small polka-dot boxfish, to the flat, beautifully hued filefish, striped angelfish, and thick schools of french grunts. As a result of severe overfishing, on many, if not most, of the world's coral reefs today there are few fish.

Things are not perfect in the Florida Keys. There are still major challenges to be addressed. Declines in water quality are still a concern, as are overfishing, coral disease, increasing incidents of coral bleaching, and damage from careless boaters and divers. The residents of the Florida Keys have recently taken a big step of their own, and central sewer and treatment systems are now under construction. For years there has been concern about the impact of poorly treated wastes seeping into nearshore waters, principally from septic systems in the region's porous limestone. For people living in the Keys, global warming is also a very serious issue. The low-lying islands of the Keys are extremely vulnerable to sea-level rise as well as the impact of hurricanes. And of course, the corals and other marine life in the region that drive tourism are also vulnerable to the effects of climate change.

Throughout the world today, only about 0.01 percent of the ocean is closed to fishing. Wild fish populations continue to decline and in some cases are on the verge of collapse. As people begin to see the benefits of marine reserves, hopefully more of the ocean will be set aside as a safe haven for fish. Ocean science and fieldwork will be essential to monitor the effectiveness of marine reserves and answer questions such as how and where best to establish them. Marine reserves alone, however, cannot solve the growing demand for fish as food.

OFFSHORE FISH FARMING

More than two billion people across the globe rely on fish as a major source of protein, and even more eat seafood as part of a healthy diet. While each day there are more human mouths to feed, there are fewer and fewer fish in the sea. To meet the growing demand for seafood, coastal fish farms are now operating across the globe, especially in China. Unfortunately, the benefits derived from these farms are often offset by their detrimental impacts. Important coastal habitats, such as mangroves or seagrass beds, are frequently destroyed to build fish farms, and concentrated wastes are released into nearby waters and can degrade water quality and cause algal blooms.

Marine biologist and aquaculture scientist Dan Benetti at the Rosenstiel School is a pioneer in the field of offshore fish farming. For years he has been studying and developing technology to efficiently and safely farm fish in the open ocean, where habitats don't need to be destroyed to create a growing facility and the sea's natural flow helps keep the fish healthy and disperse potentially harmful wastes. Dan and his colleagues have been testing pen structures, management techniques, and the environmental impacts of growing fish far from shore. Their goal is to have the highest yield with the lowest impact. The diamond-shaped cages or geodesic-domed pens they use look a bit like undersea flying saucers, and can hold up to 100,000 fish at a time. They have had great success farming snapper and cobia, and methods to raise yellowtail jacks, amberjacks, and pompano are in development. With the technology involved advancing rapidly, Dan believes that open-ocean farming of tuna is just around the corner. This is distinctly different from the tuna ranching already occurring in places such as the Mediterranean, where thousands of bluefin tuna are being captured and then fattened up in pens. True aquaculture entails raising a brood stock and farming the fish from larvae to adult, thereby forgoing wild capture.

Dan notes an interesting glitch his team encountered in their research and development. The stronger-than-Kevlar synthetic fiber netting used in their pens turned out to be shark-resistant rather than shark-proof—big difference. Sharks kept breaking into the pens to feed on injured, sick, or dead fish. Interestingly, the sharks never went inside after healthy fish. Dan's team solved the problem by improving the cage design, developing more rigid "shark-proof" netting, and regular maintenance—each day a diver removes any ailing or dead fish from the bottom of the cages.

One of the problematic issues in marine aquaculture has long been the ecological inefficiency of using wild-caught fish to feed those in a farm. Dan's group and others are working hard to move away from food based principally on wild-caught fishmeal. He is extremely optimistic about their progress and the potential of offshore aquaculture to help feed the world's masses, while simultaneously taking pressure off the ocean and its wild stocks.

With growing interest, offshore aquaculture has the potential to soon become a multibillion-dollar industry. In the United States much of the seafood we consume is imported, contributing to a rising seafood trade deficit, now estimated at about $8 billion. Dan believes that if the nation does not come up with a reasonable regulatory and permitting structure for offshore aquaculture soon, investors are going to go abroad—in fact, they already are—and the trade deficit will only increase. In March 2007 NOAA and the U.S. Department of Commerce sent a National Offshore Aquaculture Act to Congress "to create a regulatory framework that allows for safe and sustainable aquaculture operations for fish and shellfish in U.S. federal waters, three or more miles off the coast." Dan supports the act, believing it would be good for the country and the ocean, providing the permitting process needed, environmental oversight, incentives, and funds for research.

⁓⁓⁓⁓⁓⁓⁓ The ocean is changing and so, too, must our use of the sea. While the news from around the globe is alarming, there is reason for hope. Through science we are gaining an understanding of what is causing harm to the sea. Researchers must now enhance their efforts in the field and work alongside resource managers, regulatory agencies, local communities, industry, investors, and our political leaders to find solutions, implement them, monitor their effectiveness, and, hopefully, restore and protect the ocean for the future. Time is of the essence.

EPILOGUE

～～～～～～～～～ Going into the field is an exciting, challenging, and inspiring part of doing ocean science; it is also essential. In the field we gain scientific insights that cannot come from the laboratory or a computer. We observe the wonders of nature or its powerful forces, which both teach and humble us. By overcoming obstacles, we grow as scientists and as people, and often make discoveries or technical advances. Our success is built on the work of our predecessors, and tests our ingenuity and perseverance. Though many think of science as a boring, methodical process, in the field humor, creativity, and adventure come into play. And more than anything else, being in the field offers us the chance to observe and appreciate the complexities and beauty of nature firsthand.

Unfortunately, today fewer and fewer scientists and students are getting the opportunity to experience and study the sea firsthand. Funds for fieldwork and field trips are limited, and greater emphasis is being placed on remotely observed data and sophisticated computer modeling. Graduate students and scientists may now spend months investigating the ocean and never put a toe in the surf or go to sea for a week or even a day. Marine biologist Felicia Colman from Florida State University is concerned: "By relying too heavily on technology, we risk losing our connection to natural history and behavior—the very things that fascinated us as children, and drew us to science as adults."

In marine biology the new wave of students is eager and excited about molecular and genetic techniques, and rightly so. But do we want scientists who can identify an organism based solely on its DNA, yet have never seen the creature in the field or studied how it interacts with other animals and the environment? My geologist colleagues tell me that students coming into graduate school are more apt to want to study seismic lines or a three-dimensional model on a computer, rather than get dirty and dig into the mud and rocks with their bare hands. We all worry that very soon we won't have the expertise needed in disciplines such as taxonomy, ecology, or field-based geology.

Today there is justifiably a greater emphasis on applied science in the oceans to meet societal needs while less interest in basic research to understand or explore the marine environment. Unquestionably, we need a more comprehensive understanding of our impacts on the ocean and to find ways to minimize and prevent them. We must also protect people from the hazards posed by ocean-related phenomenon such as hurricanes, storm surge, and tsunamis. In the process, however, we need to ensure that we don't risk losing the understanding or unexpected benefits that come from basic and field-oriented science.

Fieldwork is also costly, and facilities to support access to the ocean are expensive to build, operate, and especially to maintain. Marine laboratories that provide access to the field are becoming fewer, and those in existence are struggling. Even the world's only operating undersea research station, a concept that was once the focus of so much worldwide attention and investment, is now fighting to survive. As opportunities and support to go into the field diminish, we also risk losing the inspiration and sense of discovery they avail.

It's not just scientists and graduate students who are not going into the field and learning about nature firsthand. Author Richard Louv powerfully showcases the problem in his book *Last Child in the Woods*, or what he calls "no child left inside." He explains that today children have fewer opportunities for unstructured play in the natural environment, and that it is contributing to childhood obesity, attention-deficit syndrome, depression, and a lack of awareness of and appreciation for nature. Children and students of all ages also have fewer opportunities to go on field trips to nature centers, to parks, and on boats, due to the cost and the concerns over liability and risk. For a scientist these trends are extremely worrisome.

My colleagues unanimously agree that the opportunity to experience nature firsthand as a child strongly influenced their decision to become a scientist and to study the ocean. Whether it was investigating tide pools with their family, running wild in the woods, or taking part in a class trip, early experiences in the field had a profound influence on their lives. Marine scientist Elizabeth Gladfelter documented this importance in a series of interviews published in the book *Agassiz's Legacy: Scientists' Reflections on the Value of Field Experience.* As fewer wild places are left and children have fewer opportunities to explore those that remain, what does this mean for the future generation of natural scientists? Experiences in nature also make us aware and appreciative of its beauty, mystery, and complexity. Without opportunities to gain this appreciation, are we also at risk of, or have we already lost, humanity's stewardship ethic toward the environment?

It is critical that scientists continue to explore and study the ocean firsthand, but it is just as essential that children and students be given the opportunity as well. With time in, on, and under the sea, who knows what ocean mysteries will be unveiled next, what problems will be solved, or what crucial piece of understanding will be obtained? To find out, we must continue to go into the field and chase science at sea.

RELATED WEB SITES

~~~~~~~~~~~~~~~~~~~~ The following Web sites, listed in alphabetical order, provide detailed information about the institutions, agencies, and organizations mentioned in the book. To contact the people referred to in the book or to learn more about their research, go to their institution's Web site.

AQUARIUS UNDERSEA HABITAT
   (www.uncw.edu/aquarius)

CALIFORNIA STATE UNIVERSITY, NORTHRIDGE
   (www.csun.edu)

CHARLES DARWIN RESEARCH STATION
   (www.darwinfoundation.org)

COASTAL DATA INFORMATION PROGRAM
   (cdip.ucsd.edu)

COASTAL STUDIES INSTITUTE, LOUISIANA STATE UNIVERSITY
   (www.csi.lsu.edu)

DALHOUSIE UNIVERSITY
   (marine.biology.dal.ca)

DUKE UNIVERSITY MARINE LAB
   (www.nicholas.duke.edu)

ECOMAR
   (espanol.geocities.com/ongprogramaecomar)

EVERGLADES NATIONAL PARK
   (www.nps.gov/ever/)

FLORIDA FISH AND WILDLIFE RESEARCH INSTITUTE
(research.myfwc.com)

FLORIDA KEYS NATIONAL MARINE SANCTUARY
(floridakeys.noaa.gov)

FLORIDA STATE UNIVERSITY
(www.fsu.edu)

FLOWER GARDEN BANKS NATIONAL MARINE SANCTUARY
(flowergarden.noaa.gov)

GALÁPAGOS NATIONAL PARK SERVICE
(www.galapagosonline.com)

HARBOR BRANCH OCEANOGRAPHIC INSTITUTION
(www.hboi.edu)

HARVARD UNIVERSITY
(www.harvard.edu)

HUBBS-SEAWORLD RESEARCH INSTITUTE
(www.hswri.org)

MAINE MARITIME ACADEMY
(www.mainemaritime.edu)

MASSACHUSETTS INSTITUTE OF TECHNOLOGY
(www.mit.edu)

MONTEREY BAY AQUARIUM
(www.mbayaq.org)

MONTEREY BAY AQUARIUM RESEARCH INSTITUTE
(www.mbari.org)

MONTEREY BAY NATIONAL MARINE SANCTUARY
(montereybay.noaa.gov)

MOTE MARINE LABORATORY
(www.mote.org)

NATIONAL AERONAUTICS AND SPACE ADMINISTRATION
(www.nasa.gov)

NATIONAL HURRICANE CENTER
(www.nhc.noaa.gov)

NATIONAL MARINE SANCTUARY PROGRAM
(sanctuaries.noaa.gov)

NATIONAL OCEANIC AND ATMOSPHERIC ADMINISTRATION
(www.noaa.gov)

NATIONAL SCIENCE FOUNDATION
(www.nsf.gov)

NOVA SOUTHEASTERN UNIVERSITY OCEANOGRAPHIC CENTER
(www.nova.edu/ocean)

OBERLIN COLLEGE
(www.oberlin.edu)

OCEAN RESEARCH AND CONSERVATION ASSOCIATION
(www.oceanrecon.org)

OCEANOGRAPHER OF THE NAVY
(www.oceanographer.navy.mil)

ROYAL DUTCH SHELL
(www.shell.com)

SCRIPPS INSTITUTION OF OCEANOGRAPHY
(sio.ucsd.edu)

SEA DWELLERS DIVE CENTER
(www.sea-dwellers.com)

SEA EDUCATION ASSOCIATION
(www.sea.edu)

SOUTHERN CALIFORNIA COASTAL WATER RESEARCH PROJECT
AUTHORITY
(www.sccwrp.org)

STANFORD UNIVERSITY
(www.stanford.edu)

STELLWAGEN BANK NATIONAL MARINE SANCTUARY
(stellwagen.noaa.gov)

TRINITY UNIVERSITY
(www.trinity.edu)

UNIVERSITY OF HAWAII
(www.soest.hawaii.edu)

UNIVERSITY OF MIAMI ROSENSTIEL SCHOOL OF MARINE AND
ATMOSPHERIC SCIENCE
(www.rsmas.miami.edu)

UNIVERSITY OF NORTH CAROLINA AT CHAPEL HILL
(www.marine.unc.edu)

UNIVERSITY OF NORTH CAROLINA AT WILMINGTON
(www.uncw.edu)

UNIVERSITY OF SOUTH FLORIDA
(www.marine.usf.edu)

UNIVERSITY OF SOUTHERN CALIFORNIA TSUNAMI RESEARCH CENTER
(www.usc.edu/dept/tsunamis)

U.S. GEOLOGICAL SURVEY
(www.usgs.gov)

U.S. OFFICE OF NAVAL RESEARCH
(www.onr.navy.mil)

WESLEYAN UNIVERSITY
  (www.wesleyan.edu)

WOODS HOLE OCEANOGRAPHIC INSTITUTION
  (www.whoi.edu)

RECOMMENDED AND RELATED READING

Ballard, R. D., and W. Hively. *The Eternal Darkness: A Personal History of Deep-Sea Exploration*. Princeton, NJ: Princeton University Press, 2002.

Benetti, D., L. Brand, J. Collins, R. Orhun, A. Benetti, B. O'Hanlon, A. Danylchuk, D. Alston, J. Rivera, and A. Cabarcas. "Can Offshore Aquaculture of Carnivorous Fish Be Sustainable? Case Studies from the Caribbean." *World Aquaculture* 37 (2006): 44–47.

Boustany, A. M., S. F. Davis, P. Pyle, S. D. Anderson, B. J. Le Boeuf, and B. A. Block. "Satellite Tagging: Expanded Niche for White Sharks." *Nature* 415 (2002): 35–36.

Broad, W. J. *The Universe Below: Discovering the Secrets of the Deep Sea*. New York: Touchstone / Simon and Schuster, 1997.

Carpenter, R. "Mass Mortality of *Diadema antillarum*." *Marine Biology* 104, no. 1 (1990): 67–77.

Chase, G. A. *Auxiliary Sail Vessel Operations*. Centreville, MD: Cornell Maritime Press, 1997.

Corson, T. *The Secret Life of Lobsters: How Fishermen and Scientists Are Unraveling the Mysteries of Our Favorite Crustacean*. New York: Harper Perennial, 2005.

Cowen, R. K. "Large Scale Pattern of Recruitment by the Labrid, *Semicossyphus pulcher*: Causes and Implications." *Journal of Marine Research* 43, no. 3 (1985): 719–42.

Cowen, R. K., C. B. Paris, and A. Srinivasan. "Scaling of Connectivity in Marine Populations." *Science* 311, no. 5760 (2006): 522–27.

Crowder, L. B., G. Osherenko, O. R. Young, S. Airamé, E. A. Norse, N. Baron, J. C. Day, F. Douvere, C. N. Ehler, B. S. Halpern, S. J. Langdon, K. L. McLeod, J. C. Ogden, R. E. Peach, A. A. Rosenberg, and J. A. Wilson.

"Sustainability: Resolving Mismatches in U.S. Ocean Governance." *Science* 313, no. 5787 (2006): 617–18.

Dean, C. *Against the Tide.* New York: Columbia University Press, 2001.

De Roy, T. *Galapagos: Islands Born of Fire.* Lynchburg, VA: Warwick House, 2000.

Earle, S. *Sea Change: A Message of the Oceans.* New York: G. P. Putnam, 1995.

Ellis, R. *The Empty Ocean.* Washington, DC: Island Press, 2003.

Feely, R. A., C. L. Sabine, K. Lee, W. Berelson, J. Kleypas, V. J. Fabry, and F. J. Millero. "Impact of Anthropogenic CO_2 on the $CaCO_3$ System in the Oceans." *Science* 305, no. 5682 (2004): 362–66.

Feingold, J. S. "Coral Survivors of the 1982–83 El Niño–Southern Oscillation, Galápagos Islands, Ecuador." *Coral Reefs* 15 (1996): 108.

———. "Responses of Three Coral Communities to the 1997–98 El Niño–Southern Oscillation: Galápagos Islands, Ecuador." *Bulletin of Marine Science* 69, no. 1 (2001): 61–77.

Feldheim, K. A., S. H. Gruber, and M. V. Ashley. "The Breeding Biology of Lemon Sharks at a Tropical Nursery Lagoon." *Proceedings of the Royal Society* (London) 269, no. 1501 (2002): 1655–61.

Forman, W. *The History of American Deep Submersible Operations.* Flagstaff, AZ: Best Publishing Co., 1999.

Gladfelter, E. H. *Agassiz's Legacy: Scientists' Reflections on the Value of Field Experience.* New York: Oxford University Press, 2002.

Glover, L., ed. *Defying Ocean's End: An Agenda for Action.* Washington, DC: Island Press, 2004.

Glynn, P. W. "Coral Reef Bleaching: Facts, Hypotheses, and Implications." *Global Change Biology* 2, no. 6 (1996): 495–509.

Gruber, S. H. "Role of the Lemon Shark, *Negaprion brevirostris* (Poey) as a Predator in the Tropical Marine Environment: A Multidisciplinary Study." *Florida Scientist* 45, no. 1 (1982): 46–75.

Halley, R. B., and E. J. Prager. *Sedimentation, Sea-Level Rise, and Circulation in Florida Bay.* USGS Fact Sheet, FS-156-96. 1996.

Halpern, B. S. "The Impact of Marine Reserves: Do Reserves Work and Does Reserve Size Matter?" *Ecological Applications* 13, no. 1, supplement (2003): S117–37.

Helvarg, D. *Blue Frontier: Dispatches from America's Ocean Wilderness.* San Francisco: Sierra Club Books, 2006.

Heupel, M. R., and R. E. Hueter. "Importance of Prey Density in Relation to the Movement Patterns of Juvenile Blacktip Sharks (*Carcharhinus limbatus*) within a Coastal Nursery Area." *Marine and Freshwater Research* 53, no. 2 (2002): 543–50.

Heupel, M. R., C. A. Simpfendorfer, and R. E. Hueter. "Running before the Storm: Blacktip Sharks Respond to Falling Barometric Pressure

Associated with Tropical Storm Gabrielle." *Journal of Fish Biology* 63, no. 5 (2003): 1357–63.

Humphris, S. *Seafloor Hydrothermal Systems: Physical, Chemical, Biological, and Geological Interactions.* Geophysical Monograph 91. Washington, DC: American Geophysical Union, 1995.

Kurlansky, M. *Cod: A Biography of the Fish That Changed the World.* New York: Vintage, 1998.

Leichter, J. J., H. L. Stewart, and S. L. Miller. "Episodic Nutrient Transport to Florida Coral Reefs." *Limnology and Oceanography* 48, no. 4 (2003): 1394–1407.

Liu, P. L-F., P. Lynett, H. Fernando, B. E. Jaffe, H. Fritz, B. Higman, R. Morton, J. Goff, and C. Synolakis. "Observations by the International Tsunami Survey Team in Sri Lanka." *Science* 308, no. 5728 (2005): 1595.

Louv, R. *Last Child in the Woods: Saving Our Children from Nature-Deficit Disorder.* Chapel Hill, NC: Algonquin Books, 2006.

McPherson, B. F., and R. Halley. *The South Florida Environment: A Region under Stress.* USGS Circular 1134. 1997.

Messing, C. G., M. C. RoseSmyth, S. R. Mailer, and J. E. Miller. "Relocation Movement in a Stalked Crinoid (Echinodermata)." *Bulletin of Marine Science* 42, no. 3 (1988): 480–87.

Myers, R. A., J. K. Baum, T. D. Shepherd, S. P. Powers, and C. H. Peterson. "Cascading Effects of the Loss of Apex Predatory Sharks from a Coastal Ocean." *Science* 315, no. 5820 (2007): 1846–50.

National Research Council. *From Monsoons to Microbes: Understanding the Ocean's Role in Human Health.* Washington, DC: National Academy Press, 1999.

Norton, T. *Underwater to Get Out of the Rain: A Love Affair with the Sea.* Cambridge, MA: De Capo Press, 2006.

Nouvian, C. *The Deep: The Extraordinary Creatures of the Abyss.* Chicago: University of Chicago Press, 2007.

Paine, R. T., and R. L. Vadas. "The Effects of Grazing by Sea Urchins, *Strongylocentrotus* spp., on Benthic Algal Populations." *Limnology and Oceanography* 14, no. 5 (1969): 710–19.

Pauly, D., V. S. Christensen, J. Dalsgaard, R. Froese, and F. Torres, Jr. "Fishing down Marine Food Webs." *Science* 279, no. 5352 (1998): 860–63.

Pomponi, S. A. "The Bioprocess-Technological Potential of the Sea." *Journal of Biotechnology* 70 (1999): 5–13.

Prager, E. J. *Furious Earth: The Science and Nature of Earthquakes, Volcanoes, and Tsunamis.* New York: McGraw-Hill, 2000.

———. *The Oceans.* New York: McGraw-Hill, 2000.

Prager, E. J., and R. N. Ginsburg. "Processes of Carbonate Nodule Growth on Florida's Outer Shelf and Its Implications for Fossil Interpretations." *Palaios* 4 (1989): 310–17.

Prager, E. J., and R. B. Halley. *Bottom Type Map of Florida Bay.* USGS Open File Report 97-526. 1997.

———. "The Influence of Seagrass on Shell Layers and Florida Bay Mudbanks." *Journal of Coastal Research* 15, no. 4 (1999): 1151–62.

Prager, E. J., J. B. Southard, and E. R. Vivoni-Gallart. "Experiments on the Entrainment Threshold in Well-sorted and Poorly Sorted Carbonate Sands." *Sedimentology* 43 (1996): 33–40.

Revkin, A. *The North Pole Was Here: Puzzles and Perils at the Top of the World.* Boston: Kingfisher, 2006.

Roberts, C. M., J. A. Bohnsack, F. Gell, J. P. Hawkins, and R. Goodridge. "Effects of Marine Reserves on Adjacent Fisheries." *Science* 294, no. 5548 (2001): 1920–23.

Rogers, C. S. "Responses of Coral Reefs and Reef Organisms to Sedimentation." *Marine Ecology Progress Series* 62 (1990): 185–202.

Rose, C. D., W. C. Sharp, W. J. Kenworthy, J. H. Hunt, W. G. Lyons, E. J. Prager, J. F. Valentine, M. O. Hall, P. E. Whitfield, and J. W. Fourqurean. "Overgrazing of a Large Seagrass Bed by the Sea Urchin *Lytechinus variegatus* in Outer Florida Bay." *Marine Ecology Progress Series* 190 (1999): 211–22.

Safina, C. *Song for the Blue Ocean: Encounters along the World's Coasts and beneath the Seas.* New York: Henry Holt, 1997.

———. *Voyage of the Turtle: In Pursuit of the Earth's Last Dinosaur.* New York: Owl Books, 2006.

Schmale, M. "Prevalence and Distribution Patterns of Tumors in Bicolor Damselfish (*Pomacentrus partitus*) on South Florida Reefs." *Marine Biology* 109, no. 2 (1991): 203–12.

Seymour, R. J. "Effects of El Niños on the West Coast Wave Climate." *Shore and Beach* 66, no. 3 (1998): 3–6.

Shinn, E. A. "Submarine Lithification of Holocene Carbonate Sediments in the Persian Gulf." *Sedimentology* 12, nos. 1–2 (1969): 109–44.

Shinn, E. A., R. M. Lloyd, and R. N. Ginsburg. "Anatomy of a Modern Carbonate Tidal-Flat, Andros Island, Bahamas." *Journal of Sedimentary Research* 39, no. 3 (1969): 1202–28.

Steneck, R. S., M. H. Graham, B. J. Bourque, D. Corbett, J. M. Erlandson, J. A. Estes, and M. J. Tegner. "Kelp Forest Ecosystems: Biodiversity, Stability, Resilience, and Future." *Environmental Conservation* 29 (2002): 436–59.

Stewart, P. D. *Galapagos: The Islands that Changed the World.* New Haven, CT: Yale University Press, 2006.

Waller, G., ed. *Sea Life: A Complete Guide to the Marine Environment.* Washington, DC: Smithsonian Institution Press, 1996.

Widder, E. A. "Bioluminescence." *Sea Technology* 38, no. 3 (1997): 33–39

Wilson, E. O. *The Future of Life.* New York: Knopf, 2002.

———. *Naturalist.* New York: Warner Books, 1995.

SPONSORS AND PARTNERING ORGANIZATIONS

The **Rosenstiel School of Marine and Atmospheric Science** at the University of Miami is one of the premier oceanographic research and education institutions in the world. Based at a sixteen-acre campus on Virginia Key in Miami, Florida, the Rosenstiel School has more than sixty years of experience in ocean research, having begun as a marine biology and fisheries laboratory in 1943. Today it is a world leader, boasting a broad global research and education agenda in areas such as ocean and atmospheric circulation, sea-level and climate change, coral reefs, fisheries, aquaculture, hurricane and monsoon mechanics, computer modeling, and ocean policy.

The **National Marine Sanctuary Foundation** (NMSF) was created to assist the federally managed National Marine Sanctuary Program with conservation-based research, education, and outreach. Their programs are designed to preserve, protect, and promote meaningful opportunities for public interaction within the nation's marine sanctuaries—our underwater treasures. NMSF is a private, nonprofit, 501(c)(3) tax-exempt organization that creates public and private-sector partnerships as needed to accomplish its mission.

The **National Marine Sanctuary Program** serves as the trustee for a system of fourteen marine protected areas encompassing more than 150,000 square miles of marine and Great Lakes waters from Washington State to the Florida Keys, and from Lake Huron to American Samoa. The system includes thirteen national marine sanctuaries and the Northwestern Hawaiian Islands Marine National Monument. The National Marine Sanctuary Program is part of the National Oceanic and Atmospheric Administration, which manages sanctuaries by working cooperatively with the public to protect sanctuaries while allowing compatible recreation and commercial activities. The sanctuary program also works to enhance public awareness of our marine resources and marine heritage through scientific research, monitoring, exploration, educational programs, and outreach.

The **Wildlife Foundation of Florida** is preeminently an advocate for Florida's fish and wildlife resources. A not-for-profit, nonpolitical, financing and action group, they are dedicated to preserving Florida's fish and wildlife for future generations. Now in the second decade of service, the foundation functions to raise funds and to build support for the Florida Fish and Wildlife Conservation Commission and other organizations engaged in science-based nature conservation, management, education, and research activities.

The **Fish and Wildlife Research Institute** is part of the Florida Fish and Wildlife Conservation Commission. The institute's statewide research programs focus on obtaining data and information needed by natural resource managers and stakeholders, including assessment and restoration of ecosystems and studies of freshwater and marine fisheries, aquatic and terrestrial wildlife, imperiled species, and red tides. The institute also engages in outreach activities to complement all of its programs.